THE OXFORD BOOK OF COMIC VERSE

THE

Oxford Book of

Comic Verse

❖

Edited by

JOHN GROSS

Oxford New York
OXFORD UNIVERSITY PRESS
1994

Oxford University Press, Walton Street, Oxford OX2 6DP

Oxford New York
Athens Auckland Bangkok Bombay
Calcutta Cape Town Dar es Salaam Delhi
Florence Hong Kong Istanbul Karachi
Kuala Lumpur Madras Madrid Melbourne
Mexico City Nairobi Paris Singapore
Taipei Tokyo Toronto

and associated companies in
Berlin Ibadan

Oxford is a trade mark of Oxford University Press

First published 1994

British Library Cataloguing in Publication Data
Data available

Library of Congress Cataloging-in-Publication Data
The Oxford book of comic verse / edited by John Gross.
p. cm.
Includes bibliographical references (p.) and index.
1. Humorous poetry, English. 2. Humorous poetry, American.
I. Gross, John J.
PR1195.H8088 1994 821'.0708—dc20 94-656
ISBN 0-19-214207-0

1 3 5 7 9 10 8 6 4 2

Typeset by Wyvern Typesetting Ltd, Bristol
Printed in Great Britain
on acid-free paper at
The Bath Press
Bath, Avon

Contents

Introduction

Comic verse is verse that is designed to amuse—and perhaps that is as far as any attempt at a definition ought to go. One step further, and you plunge into a tangle of qualifications, distinctions, discriminations, boundary-disputes. Who, after all, would want to undertake a detailed definition of comedy itself? There is high comedy and low comedy, light comedy and dark comedy, learned comedy and popular comedy. There are also fifty-seven varieties of comic verse, and nine and sixty ways of composing it.

My own approach in compiling the present anthology has been firmly (or, I suppose some might say, weakly) eclectic. When it comes to deciding whether an individual poem qualifies for admission, I have worked by instinct rather than rule. I have not only accepted the multifariousness of the genre as a whole, but welcomed it; and any impression of variety has (I hope) been reinforced by an attempt to accord proper recognition to writers from outside the British Isles—from America above all, but from other parts of the English-speaking world as well.

Even an eclectic anthologist has to draw the line somewhere, however, and there are bound to be debatable decisions. In the case of comic verse, one notably problematic area is satire. As an expression of hatred or contempt, it carries us too far beyond comedy for comfort. As a form of ridicule, on the other hand, it shades into the more general category of what might be called 'comic verse with a sting'. Genial satire, in short, is a variety of comic verse, savage satire isn't; satire which takes its chief pleasure in contemplating its target qualifies, satire which takes its chief pleasure in demolishing its target doesn't. But then what happens when a satirist changes course in mid-poem? One of the items I was originally determined to include in this collection was the account of the city merchant Balaam, from the third of Alexander Pope's *Moral Essays* (the Epistle to Bathurst, 'Of the Use of Riches'). Whatever moral we are meant to draw from this particular essay, the early stages of Balaam's social ascent are pure comedy:

> Sir Balaam now, he lives like other folks,
> He takes his chirping pint, and cracks his jokes:
> 'Live like yourself,' was soon my Lady's word;
> And lo! two puddings smoked upon the board.

Magnificent. But as things go wrong for Balaam, the mood of the piece grows harsher, until finally we find ourselves confronted with pure venom:

His daughter flaunts a Viscount's tawdry wife;
She bears a Coronet and Pox for life.

I tried hard, I very nearly persuaded myself, but in the end I just couldn't find a place in an anthology of comic verse for *that*. Nor could I see any way of breaking off earlier in the story, without seriously violating Pope's artistry. So the whole passage, alas, had to go.

In its way, the problem of satire is simply one aspect of a more general question: how comic does a comic poem need to be? It would be possible, no doubt, to compile an anthology consisting of nothing but poems proclaiming 'I'm funny. Please laugh'—it would be possible, but not very satisfactory. A diet of uninterrupted merriment soon palls; it also presupposes a depressingly rigid view of comedy (and beyond that, of human nature). No, an adequate Book of Comic Verse must always be to some extent a Book of Serio-Comic Verse—with the proviso that even in its more serio moments, the comic must always predominate.

It would be equally wrong to assume that comic verse is necessarily a minor art. Chaucer alone is enough to disprove that: a great poet, never greater than when he is expounding an essentially comic vision of life. The same could be said of Byron, Burns, arguably of Dryden.

Yet one shouldn't pitch one's claims too high. The two wittiest poets in the language, for instance, are Donne and Marvell—or so I would argue; but it would be quite inappropriate, even demeaning, to affix the label 'comic verse' to 'The Canonization', shall we say, or 'To His Coy Mistress'. The wit of such poetry belongs to a finer, more intricate, more serious tradition. In fact there is nothing by either man in this anthology: those of their poems which *could* reasonably be classified as comic verse—Donne's epigrams, for instance or Marvell's *The Character of Holland*—seem to me to fall too far below their best work to do them justice.

The 'comic' label suggests limitations, then. In spite of a Chaucer or a Byron, comic verse as a whole does undoubtedly constitute one of the secondary departments of literature. That, indeed, is part of its charm. It is undemanding; it sticks to the accessible foothills.

English is rich not only in comic verse, but in comic-verse specialists—in what might be called professionals. Matthew Prior; John Wolcot *alias* 'Peter Pindar'; a golden succession of comic poets in the nineteenth century; Belloc and Chesterton; lesser stars like E. C. Bentley of the clerihews and Harry Graham of *Ruthless Rhymes*—does any other literature have more to show in this particular line? And while it is true that after its Victorian and Edwardian heyday the professional tradition ran thin, and frequently degenerated into mere whimsy (or, as somebody said, '*Punch* without punch'), there have been powerful modern compensations. For the past sixty years or so, most of the best comic verse has been the work of 'serious' poets rather than humorists—of writers like Betjeman and Stevie Smith, Ewart and Larkin. (In America, although the situation is now much the same, the professional tradition held its ground longer. At least two outstanding practitioners, Ogden Nash and Phyllis McGinley, were still in business well after the Second World War.)

Comic verse has its masters and masterpieces. It also has its oddities, its happy accidents, its moments of inspired amateurism. If it can give rise to feats of dazzling verbal and metrical virtuosity, it can also come dangerously close to doggerel—and still work. In certain moods, it seems a game that anyone can play; and although there are reams of rhymed ineptitude to testify that this isn't so, it remains—more than any other branch of verse—the realm of the casual and the informal, the quirky and the miscellaneous. Praed and Hood and W. S. Gilbert (to say nothing of Chaucer and Byron) ought to come first, but an anthology which aims to take the true measure of the territory also ought to find at least some space for the anonymous limerick, the weekend competition, the advertising jingle, the marching song, the music-hall monologue, the little piece of nonsense that you can't get out of your head.

I have left one large source of material untapped, however. Perhaps no form of comic verse enjoys such genuine popularity as the bawdy lyric—I mean the fully fledged four-letter variety. Few of us can have gone through life without at some stage encountering Mr Banglestein or Eskimo Nell or the monk of great renown; the Ball of Kirriemuir and the Good Ship Venus are an undoubted part of our heritage. None the less I decided, after due consideration, not to include the songs in which they figure, or anything else of the same kind. This is mainly because I think it spoils such things to see them set down in print. Their whole point is that they should remain half-hidden, part of an underground tradition. But I must admit that I also find the more extreme (and presumably more authentic) versions of some 'unprintable' songs disgusting. At one stage, for example, I contemplated including 'The One-Eyed Riley'. It is a song that has been sanctified, for literary purposes, by being quoted by Eliot in *The Cocktail Party*, and such snatches of it as I had come across elsewhere made it sound amusing. But it was a tidied-up version that I had been overhearing, and when I began trying to track down the genuine article—in such works as Richard Slatta's learned study of cowboy ballads, *All the Whorehouse Bells Were Ringing* (1993)—I soon changed my mind.

A much bigger question was deciding what to do about the lyrics of popular songs. Most lyrics are nothing without the music, and even the best of them lose a good deal. If, on the other hand, you find yourself recalling the music while reading them, it is liable to get in the way: verse rhythms and song rhythms can be badly at odds. So there is a problem, either way. And yet I cannot agree with Kingsley Amis when he writes that 'the wit of the witty lyric as written by Cole Porter evaporates as soon as it comes out of the end of the transcriber's pen'. My own view is that, though much is taken, much remains—more than enough to justify transcribing and reprinting not only Porter, but a number of other song-writers as well, including some of the forgotten heroes who wrote the words for music-hall songs.

In making the selection as a whole, I have tried to strike a balance between the familiar and the unexpected, the mainstream and the lesser-known. That means that some readers are going to look in vain for old favourites, and others are going to wonder how I can have wasted space on this or that chestnut. But then there are two lessons which every anthologist has to learn: he can't include everything, and he can't please everybody.

The medieval poems in the book appear in their original form. From the Tudor period onwards, beginning with Skelton, spelling and punctu-ation have been modernized. Where the titles of poems have been supplied by the editor, they are put in square brackets.

Explanatory notes have been provided only in those cases where I felt readers would be seriously held up without them. One or two of them, explaining points which don't need to be explained in Britain, are for the benefit of readers in other countries.

I would like to thank the following for help or advice of various kinds: Melanie Aspey, Tom Baistow, Jonathan Cecil, Roy Hudd, Alan Jenkins, Christopher Logue, John Julius Norwich, Elizabeth Peters, Julie Stainforth, Max Tyler, and Arline Youngman.

JOHN GROSS

THE

Oxford Book of

Comic Verse

❖

GEOFFREY CHAUCER

1340?–1400

1 from *The General Prologue to the Canterbury Tales*

(i)

[*The Monk*]

A Monk ther was, a fair for the maistrie,
An outridere, that lovede venerie,
A manly man, to been an abbot able.
Ful many a deyntee hors hadde he in stable,
And whan he rood, men myghte his brydel heere
Gynglen in a whistlynge wynd als cleere
And eek as loude as dooth the chapel belle.
Ther as this lord was kepere of the celle,
The reule of seint Maure or of seint Beneit,
By cause that it was old and somdel streit
This ilke Monk leet olde thynges pace,
And heeld after the newe world the space.
He yaf nat of that text a pulled hen,
That seith that hunters ben nat hooly men,
Ne that a monk, whan he is recchelees,
Is likned til a fissh that is waterlees,—
This is to seyn, a monk out of his cloystre.
But thilke text heeld he nat worth an oystre;
And I seyde his opinion was good.
What sholde he studie and make hymselven wood,
Upon a book in cloystre alwey to poure,
Or swynken with his handes, and laboure,
As Austyn bit? How shal the world be served?
Lat Austyn have his swynk to hym reserved!
Therfore he was a prikasour aright:
Grehoundes he hadde as swift as fowel in flight;
Of prikyng and of huntyng for the hare
Was al his lust, for no cost wolde he spare.
I seigh his sleves purfiled at the hond
With grys, and that the fyneste of a lond;

for the maistrie] pre-eminently *outridere*] a monk permitted to go outside the monastery on business *venerie*] hunting *celle*] a subordinate monastery *Beneit*] Benedict *streit*] strict *recchelees*] neglectful of his duties *what*] why *wood*] mad *swynken*] work *Austyn*] Augustine *prikasour*] hunter on horseback *purfiled*] trimmed *grys*] costly fur

And, for to festne his hood under his chyn,
He hadde of gold ywroght a ful curious pyn;
A love-knotte in the gretter ende ther was.
His heed was balled, that shoon as any glas,
And eek his face, as he hadde been enoynt.
He was a lord ful fat and in good poynt;
His eyen stepe, and rollynge in his heed,
That stemed as a forneys of a leed;
His bootes souple, his hors in greet estaat.
Now certeinly he was a fair prelaat;
He was nat pale as a forpyned goost.
A fat swan loved he best of any roost.

(ii)

[*The Summoner and the Pardoner*]

A SOMONOUR was ther with us in that place,
That hadde a fyr-reed cherubynnes face,
For saucefleem he was, with eyen narwe.
As hoot he was and lecherous as a sparwe,
With scalled browes blake and piled berd.
Of his visage children were aferd.
Ther nas quyk-silver, lytarge, ne brymstoon,
Boras, ceruce, ne oille of tartre noon;
Ne oynement that wolde clense and byte,
That hym myghte helpen of his whelkes white,
Nor of the knobbes sittynge on his chekes.
Wel loved he garleek, oynons, and eek lekes,
And for to drynken strong wyn, reed as blood;
Thanne wolde he speke and crie as he were wood.
And whan that he wel dronken hadde the wyn,
Thanne wolde he speke no word but Latyn.
A fewe termes hadde he, two or thre,
That he had lerned out of som decree—
No wonder is, he herde it al the day;
And eek ye knowen wel how that a jay
Kan clepen 'Watte' as wel as kan the pope.
But whoso koude in oother thyng hym grope,
Thanne hadde he spent al his philosophie;
Ay '*Questio quid iuris*' wolde he crie.

stepe] prominent *of a leed*] under a cauldron *forpyned*] wasted away
Somonour] an officer who delivered summonses to appear before ecclesiastical courts
saucefleem] pimpled *piled*] scanty *lytarge*] lead ointment
ceruce] white lead *grope*] test him further *Questio quid iuris*] 'the question is, what is the law on this point'

He was a gentil harlot and a kynde;
A bettre felawe sholde men noght fynde.
He wolde suffre for a quart of wyn
A good felawe to have his concubyn
A twelf month, and excuse hym atte fulle;
Ful prively a fynch eek koude he pulle.
And if he foond owher a good felawe,
He wolde techen him to have noon awe
In swich caas of the ercedekenes curs,
But if a mannes soule were in his purs;
For in his purs he sholde ypunysshed be.
'Purs is the ercedekenes helle,' seyde he.
But wel I woot he lyed right in dede;
Of cursyng oghte ech gilty man him drede,
For curs wol slee right as assoillyng savith,
And also war hym of a *Significavit*.
In daunger hadde he at his owene gise
The yonge girles of the diocise,
And knew hir conseil, and was al hir reed.
A gerland hadde he set upon his heed
As greet as it were for an ale-stake.
A bokeleer hadde he maad hym of a cake.
With hym ther rood a gentil PARDONER
Of Rouncivale, his freend and his compeer,
That streight was comen fro the court of Rome.
Ful loude he soong 'Com hider, love, to me!'
This Somonour bar to hym a stif burdoun;
Was nevere trompe of half so greet a soun.
This Pardoner hadde heer as yelow as wex,
But smothe it heeng as dooth a strike of flex;
By ounces henge his lokkes that he hadde,
And therwith he his shuldres overspradde;
But thynne it lay, by colpons oon and oon.
But hood, for jolitee, wered he noon,
For it was trussed up in his walet.
Hym thoughte he rood al of the newe jet;

harlot] rascal *ercedekenes curs*] archdeacon's excommunication
assoillyng] absolution *Significavit*] writ remanding an excommunicated person
daunger] control *reed*] adviser *ale-stake*] signpost of an inn *cake*] loaf
Pardoner] a seller of papal indulgences *Rouncivale*] a hospital near Charing Cross
burdoun] ground melody *strike of flex*] hank of flax *ounces*] small bunches
colpons] small strands *jet*] fashion

Dischevelee, save his cappe, he rood al bare.
Swiche glarynge eyen hadde he as an hare.
A vernycle hadde he sowed upon his cappe.
His walet lay biforn hym in his lappe,
Bretful of pardoun, comen from Rome al hoot.
A voys he hadde as smal as hath a goot.
No berd hadde he, ne nevere sholde have;
As smothe it was as it were late shave.
I trowe he were a geldyng or a mare.
But of his craft, fro Berwyk into Ware,
Ne was ther swich another pardoner.
For in his male he hadde a pilwe-beer,
Which that he seyde was Oure Lady veyl:
He seyde he hadde a gobet of the seyl
That Seint Peter hadde, whan that he wente
Upon the see, til Jhesu Crist hym hente.
He hadde a croys of latoun ful of stones,
And in a glas he hadde pigges bones.
But with thise relikes, whan that he fond
A povre person dwellynge upon lond,
Upon a day he gat hym moore moneye
Than that the person gat in monthes tweye;
And thus, with feyned flaterye and japes,
He made the person and the peple his apes.
But trewely to tellen atte laste,
He was in chirche a noble ecclesiaste.
Wel koude he rede a lessoun or a storie,
But alderbest he song an offertorie;
For wel he wiste, whan that song was songe,
He moste preche and wel affile his tonge
To wynne silver, as he ful wel koude;
Therefore he song the murierly and loude.

vernycle] a small copy of the handkerchief of St Veronica *Bretful*] brimful
male] bag *pilwe-beer*] pillowcase *gobet*] piece *hente*] recruited
latoun] a copper alloy *person*] parson *alderbest*] best of all *affile*] sharpen
the murierly] the more joyfully

ANONYMOUS

2

I Have a Gentle Cock

I HAVE a gentle cock,
Croweth me day:
He doth me risen erly
My matins for to say.

I have a gentle cock,
Comen he is of gret:
His comb is of red coral,
His tail is of jet.

I have a gentle cock,
Comen he is of kinde:
His comb is of red coral,
His tail is of inde.

His legges ben of asor,
So gentle and so smale;
His spores arn of silver whit
Into the wortewale.

His eynen arn of cristal,
Loken all in aumber:
And every night he percheth him
In mine ladye's chaumber.

ANONYMOUS

3

Bring us in Good Ale

BRING us in good ale, and bring us in good ale,
Fore our blessed Lady sak, bring us in good ale.

doth] makes *of gret*] of distinguished family *of kinde*] of noble stock
inde] indigo *asor*] azure *Into the wortewale*] up to the root (of the spurs)
loken] set

Bring us in no browne bred, fore that is mad of brane;
Nor bring us in no whit bred, fore therin is no game:
But bring us in good ale.

Bring us in no befe, for ther is many bones;
But bring us in good ale, for that goth downe at ones,
And bring us in good ale.

Bring us in no bacon, for that is passing fat;
But bring us in good ale, and give us inought of that,
And bring us in good ale.

Bring us in no mutton, for that is ofte lene;
Nor bring us in no tripes, for they be seldom clene:
But bring us in good ale.

Bring us in no egges, for ther ar many shelles;
But bring us in good ale, and give us nothing elles,
And bring us in good ale.

Bring us in no butter, for therin ar many heres;
Nor bring us in no pigges flesh, for that will mak us bores:
But bring us in good ale.

Bring us in no podinges, for therin is all gotes blod;
Nor bring us in no venison, for that is not for our good:
But bring us in good ale.

Bring us in no capon's flesh, for that is ofte der;
Nor bring us in no dokes flesh for they slobber in the mer:
But bring us in good ale.

mad of brane] made of bran   *podinges*] black puddings   *gotes blod*] goat's blood
der] expensive   *mer*] pond

ANONYMOUS

4

Smoke-Blackened Smiths

Swarte-smeked smethes, smatered with smoke,
Drive me to deth with den of her dintes:
Swich nois on nightes ne herd men never,
What knavene cry and clatering of knockes!
The cammede kongons cryen after 'Col! col!'
And blowen here bellewes that all here brain brestes.
'Huf, puf,' seith that on, 'Haf, paf,' that other.
They spitten and sprawlen and spellen many spelles,
They gnawen and gnacchen, they grones togidere.
And holden them hote with here hard hamers.
Of a bole hide ben here barm-felles,
Here shankes ben shakeled for the fere-flunderes.
Hevy hameres they han that hard ben handled.
Stark strokes they striken on a steled stocke.
'Lus, bus, las, das,' rowten by rowe.
Swiche dolful a dreme the Devil it todrive!
The maister longeth a litil and lasheth a lesse,
Twineth hem twein and toucheth a treble.
'Tik, tak, hic, hac, tiket, taket, tik, tak,
Lus, bus, las, das'. Swich lif they leden
Alle clothemeres, Christ hem give sorwe!
May no man for brenwateres on night han his rest.

Swarte-smeked] smoke-blackened *smatered*] begrimed *den*] din *dintes*] blows *knavene cry*] crying of men *cammede kongons*] snub-nosed changelings *Col*] coal *brestes*] bursts *gnacchen*] gnash *holden them hote*] keep themselves hot *bole hide*] bull's hide *barm-felles*] aprons *shakeled for the fere-flunderes*] protected against the sparks *that hard ben handled*] that are handled hard *steled stocke*] steel anvil *rowten by rowe*] the blows come crashing down *the Devil it todrive*] may the Devil do away with it *longeth a litil and lasheth a lesse*] stretches a small piece of iron and hammers a still smaller one *toucheth a treble*] strikes a treble note *clothemeres*] smiths who make armour for horses *for brenwateres*] because of the hissing of the water they heat up

JOHN SKELTON

1460?–1529

5 from *Philip Sparrow*

(The speaker is meant to be a young girl, Jane Scrope, a pupil at the convent school of Carrow, near Norwich)

Pla ce bo!
Who is there, who?
Di le xi!
Dame Margery.
Fa, re, my, my.
Wherefore and why, why?
For the soul of Philip Sparrow,
That was late slain at Carow,
Among the Nones Black.
For that sweet soules sake,
And for all sparrowes' souls
Set in our bede-rolls,
Pater noster qui,
With an *Ave Mari*,
And with the corner of a Creed,
The more shall be your meed.

When I remember again
How my Philip was slain,
Never half the pain
Was between you twain,
Pyramus and Thisbe,
As then befell to me.
I wept and I wailed,
The tears down hailed,
But nothing it availed
To call Philip again,
Whom Gib, our cat, hath slain.
Gib, I say, our cat
Worrowed her on that
Which I loved best.
It cannot be exprest

Placebo] the first word of the Roman Office for the Dead *Nones Black*] Benedictine nuns *bede-rolls*] lists of people to be prayed for *Worrowed*] worried

My sorrowful heaviness,
But all without redress!
For within that stound,
Half slumbering, in a sound
I fell downe to the ground.

. . . . . .

Would God I had Zenophontes,
Or Socrates the wise,
To shew me their device
Moderately to take
This sorrow that I make
For Philip Sparrow's sake!
So fervently I shake,
I feel my body quake;
So urgently I am brought
Into careful thought.
Like Andromach, Hector's wife,
Was weary of her life,
When she had lost her joy,
Noble Hector of Troy;
In like manner also
Increaseth my deadly woe,
For my sparrow is go.

It was so pretty a fool,
It would sit on a stool,
And learned after my school
For to keep his cut,
With 'Philip, keep your cut!'

It had a velvet cap,
And would sit upon my lap,
And seek after small wormes,
And sometime white bread-crummes;
And many times and oft
Between my brestes soft
It would lie and rest;
It was proper and prest.

stound] hour *Zenophontes*] probably Xenophon *keep his cut*] behave well *prest*] alert

Sometime he would gasp
When he saw a wasp;
A fly or a gnat,
He would fly at that;
And prettily he would pant
When he saw an ant,
Lord, how he would pry
After the butterfly!
Lord, how he would hop
After the gressop!
And when I said, 'Phip, Phip!'
Then he would leap and skip,
And take me by the lip.
Alas, it will me slo
That Philip is gone me fro!

. . . .

That vengeance I ask and cry,
By way of exclamation,
On all the whole nation
Of Cattes wild and tame:
God send them sorrow and shame!
That Cat specially
That slew cruelly
My little pretty sparrow
That I brought up at Carow.

O cat of carlish kind,
The fiend was in thy mind
When thou my bird untwined!
I would thou hadst been blind!
The leopardes savage,
The lions in their rage
Might catch thee in their paws,
And gnaw thee in their jaws!
The serpents of Libany
Might sting thee venomously!
The dragons with their tongues
Might poison thy liver and lungs!
The manticors of the mountains
Might feed them on thy brains!

gressop] grasshopper   *slo*] slay   *carlish*] churlish   *Libany*] Libya
manticors] fabulous beasts

Melanchates, that hound
That plucked Actaeon to the ground,
Gave him his mortal wound,
Changed to a deer,
The story doth appear,
Was changed to an hart:
So thou, foul cat that thou art,
The selfsame hound
Might thee confound,
That his own lorde bote,
Might bite asunder thy throat!

Of Ind the greedy grypes
Might tear out all thy tripes!
Of Arcady the bears
Might pluck away thine ears!
The wild wolf Lycaon
Bite asunder thy backbone!
Of Etna the burning hill,
That day and night burneth still,
Set in thy tail a blaze
That all the world may gaze
And wonder upon thee,
From Ocean the great sea
Unto the Isles of Orcady,
From Tilbury Ferry
To the plain of Salisbury!
So traitorously my bird to kill
That never ought thee evil will!

Was never bird in cage
More gentle of corage
In doing his homage
Unto his soveraine.
Alas, I say again,
Death hath departed us twain!
The false cat hath thee slain:
Farewell, Philip, adew!
Our Lord, thy soul rescue!
Farewell, without restore,
Farewell, for evermore!

bote] bit *grypes*] gryphons *Orcady*] Orkney

6 from *Speak, Parrot*

MY name is Parrot, a bird of Paradise,
  By nature devised of a wonderous kind,
Daintily dieted with divers delicate spice
  Till Euphrates, that flood, driveth me into Ind;
  Where men of that country by fortune me find
And send me to greate ladyes of estate:
Then Parrot must have an almond or a date.

A cage curiously carven, with a silver pin,
  Properly painted, to be my coverture;
A mirror of glasse, that I may toot therein:
  These maidens full meekly with many a divers flower,
  Freshly they dress, and make sweet my bower,
With 'Speak, Parrot, I pray you!' full curtesly they say,
'Parrot is a goodly bird, a pretty popegay.'

With my beak bent, my little wanton eye,
  My feathers fresh as is the emerald green,
About my neck a circulet like the rich ruby,
  My little legges, my feet both feat and clean,
  I am a minion to wait upon a queen.
'My proper Parrot, my little pretty fool!'
With ladies I learn, and go with them to school.

'Hagh! Ha! Ha! Parrot, ye can laugh prettily!'
  Parrot hath not dined all this long day.
Like your puss-cat, Parrot can mute and cry
  In Latin, Hebrew, Araby and Chaldy;
  In Greeke tongue Parrot can both speak and say,
As Persius, that poet, doth report of me,
Quis expedivit psittaco suum chaire?

Douce French of Paris Parrot can learne,
  Pronouncing my purpose after my property,
With 'Perliez byen, Parrot, ou perlez rien!'
  With Dutch, with Spanish, my tongue can agree,
  In England to God Parrot can supply:
'Christ save King Henry the Eighth, our royal king,
The red rose in honour to flourish and spring!

mute] mew *Quis expedivit . . .*] Who taught Parrot so say 'hallo'? *Perliez*] *parlez* *supply*] pray

With Katherine incomparable, our royal queen also,
That peerless pomegranate, Christ save her noble grace!'
Parrot saves habler Castiliano,
With fidasso di cosso in Turkey and in Thrace;
Vis consilii expers, as teacheth me Horace,
Mole ruit sua, whose dictates are pregnant,
Soventez foys, Parrot, en souvenante.

My lady mistress, Dame Philology,
Gave me a gifte, in my nest when I lay,
To learn all language, and it to speak aptely,
Now pandez mory, wax frantic, some men say,
Phronesis for Phrenesis may not hold her way.
An almond now for Parrot, delicately drest:
In Salve festa dies, toto there doth best.

Moderata juvant, but toto doth exceed:
Discretion is mother of noble virtues all.
Myden agan in Greeke tongue we read.
But reason and wit wanteth their provincial
When wilfulness is vicar general.
Haec res acu tangitur, Parrot, par ma foy:
Ticez-vous, Parrot, tenez-vous coy!

7 from *Colin Clout*

FOR though my rhyme be ragged,
Tattered and jagged,
Rudely rain-beaten,
Rusty and moth-eaten,
If ye take well therewith,
It hath in it some pith.

saves habler Castiliano] can speak Castilian *fidasso di cosso*] a corruption of *fidarsi in se stesso*, to trust in oneself *Vis consilii . . .*] strength without wisdom falls by its own weight *Soventez foys . . . en souvenante*] often within memory *pandez mory*] meaning unknown *Phronesis*] intelligence *Phrenesis*] frenzy *Salve festa dies*] hail, festive day *Moderata juvant*] moderation pleases *Myden agan*] nothing in excess (Greek) *Haec res . . .*] this hits the nail on the head *Ticez-vous*] *taisez-vous* *tenez-vous coy*] be quiet

ROBERT WISDOME

d. 1568

8

A Religious Use of Taking Tobacco

THE Indian weed withered quite,
Green at morn, cut down at night,
Shows thy decay;
All flesh is hay:
Thus think, then drink Tobacco.

And when the smoke ascends on high,
Think thou behold'st the vanity
Of worldly stuff,
Gone with a puff:
Thus think, then drink Tobacco.

But when the pipe grows foul within,
Think of thy soul defiled with sin.
And that the fire
Doth it require:
Thus think, then drink Tobacco.

The ashes that are left behind,
May serve to put thee still in mind
That into dust
Return thou must:
Thus think, then drink Tobacco.

JOHN LYLY

*c.*1554–1606

9

from *Endimion*

Watch STAND! Who goes there?
We charge you, appear
Fore our constable here

(In the name of the Man in the Moon).
To us billmen relate
Why you stagger so late,
And how you come drunk so soon.

Pages What are ye (scabs)?

Watch The Watch.
This the constable.

Pages A patch.

Constable Knock 'em down unless they all stand.
If any run away,
'Tis the old watchman's play,
To reach him a bill of his hand.

Pages O, gentlemen, hold,
Your gowns freeze with cold,
And your rotten teeth dance in your head.

Epiton Wine, nothing shall cost ye.

Samias Nor huge fires to roast ye.

Dares Then soberly let us be led.

Constable Come, my brown bills we'll roar,
Bounce loud at tavern door,

Omnes And i' th' morning steal all to bed.

SIR JOHN HARINGTON

1561–1612

10

TREASON doth never prosper, what's the reason?
For if it prosper, none dare call it treason.

billmen] watchmen armed with bills *patch*] fool *brown bills*] watchmen
Bounce] knock

WILLIAM SHAKESPEARE

1564–1616

11 from *The Comedy of Errors*

[*Dr Pinch*]

BY the way we met
My wife, her sister and a rabble more
Of vile confederates; along with them
They brought one Pinch, a hungry lean-faced villain;
A mere anatomy, a mountebank,
A thread-bare juggler and a fortune-teller,
A needy-hollow-eyed-sharp-looking-wretch,
A living dead man.

12 from *The Taming of the Shrew*

[*Katharina's Gown*]

Petruchio THY gown? Why, ay. Come, tailor, let us see't.
O mercy, God! What masquing stuff is here?
What's this? A sleeve? 'Tis like a demi-cannon
What, up and down, carved like an apple-tart?
Here's snip and nip and cut and slish and slash,
Like to a censer in a barber's shop.

13 from *Love's Labour's Lost*

[*Berowne in Love*]

O! AND I forsooth in love!
I, that have been love's whip;
A very beadle to a humorous sigh;
A critic, nay, a night-watch constable,
A domineering pedant o'er the boy,
Than whom no mortal so magnificent!
This wimpled, whining, purblind, wayward boy,
This signor junior, giant-dwarf, dan Cupid;
Regent of love rhymes, lord of folded arms,

12 *censer*] possibly a perfume-burner, but obscure

13 *wimpled*] blindfold

The anointed sovereign of sighs and groans,
Liege of all loiterers and malcontents,
Dread prince of plackets, king of codpieces,
Sole imperator and great general
Of trotting paritors: O my little heart!
And I to be a corporal of his field,
And wear his colours like a tumbler's hoop!
What! I love! I sue! I seek a wife!
A woman that is like a German clock,
Still a-repairing, ever out of frame,
And never going aright, being a watch,
But being watch'd that it may still go right!
Nay to be perjur'd, which is worst of all;
And among three, to love the worst of all;
A whitely wanton with a velvet brow,
With two pitch-balls stuck in her face for eyes;
Ay and by heaven, one that will do the deed
Though Argus were her eunuch and her guard:
And I to sigh for her! to watch for her!
To pray for her! Go to; it is a plague
That Cupid will impose for my neglect
Of his almighty dreadful little might.
Well, I will love, write, sigh, pray, sue, and groan:
Some men must love my lady, and some Joan.

14 from *As You Like It*

[*Jacques and Touchstone*]

A FOOL, a fool! I met a fool i' th' forest,
A motley fool: a miserable world!
As I do live by food, I met a fool,
Who laid him down and bask'd him in the sun,
And rail'd on Lady Fortune in good terms,
In good set terms, and yet a motley fool.
'Good morrow, fool', quoth I. 'No, sir', quoth he,
'Call me not fool, till heaven hath sent me fortune'.
And then he drew a dial from his poke,
And looking on it, with lack-lustre eye,
Says, very wisely, 'It is ten o'clock.
Thus we may see', quoth he, 'how the world wags:
'Tis but an hour ago since it was nine,

paritors] officers of the Ecclesiastical Courts

And after one hour more 'twill be eleven;
And so from hour to hour, we ripe, and ripe,
And then from hour to hour, we rot, and rot,
And thereby hangs a tale.' When I did hear
The motley fool thus moral on the time,
My lungs began to crow like chanticleer,
That fools should be so deep-contemplative;
And I did laugh, sans intermission,
An hour by his dial. O noble fool!
A worthy fool! Motley's the only wear.

15 from *The Tempest*

[*A Sea Song*]

THE master, the swabber, the boatswain and I,
  The gunner and his mate,
Loved Mall, Meg, and Marian and Margery,
  But none of us cared for Kate;
  For she had a tongue with a tang,
  Would cry to a sailor, 'Go hang!'
She loved not the savour of tar nor of pitch,
Yet a tailor might scratch her where'er she did itch:
  Then to sea, boys, and let her go hang.

JOHN DAVIES OF HEREFORD

1565?–1618

16 *The author loving these homely meats specially, viz.: cream, pancakes, buttered pippin-pies (laugh, good people) and tobacco; writ to that worthy and virtuous gentlewoman, whom he calleth mistress, as followeth*

IF there were, oh! an Hellespont of cream
Between us, milk-white mistress, I would swim
To you, to show to both my love's extreme,
Leander-like,—yea! dive from brim to brim.

But met I with a buttered pippin-pie
Floating upon 't, that would I make my boat
To waft me to you without jeopardy,
Though sea-sick I might be while it did float.
Yet if a storm should rise, by night or day,
Of sugar-snows and hail of caraways,
Then, if I found a pancake in my way,
It like a plank should bring me to your kays;
  Which having found, if they tobacco kept,
  The smoke should dry me well before I slept.

SIR JOHN DAVIES

1569–1626

17

Francus

WHEN Francus comes to solace with his whore
He sends for rods and strips himself stark naked,
For his lust sleeps and will not rise before
By whipping of the wench it be awaked.
I envy him not, but wish I had the power
To make myself his wench but one half hour.

SAMUEL ROWLANDS

1570?–1630?

18

Sir Revel

'SPEAK, gentlemen, what shall we do today?
Drink some brave health upon the Dutch carouse?
Or shall we to the Globe and see a play?
Or visit Shoreditch for a bawdy-house?
Let's call for cards or dice, and have a game:
To sit thus idle is both sin and shame.'

This speaks Sir Revel, furnished out with fashion,
From dish-crowned hat unto the shoes' square toe,
That haunts a whore-house but for recreation,
Plays but at dice to coney-catch or so;
Drinks drunk in kindness, for good fellowship,
Or to the play goes but some purse to nip.

BEN JONSON

1573?–1637

19

Comus's Song

ROOM, room, make room for the bouncing belly!
First father of sauce, and deviser of jelly:
Prime master of arts and the giver of wit,
That found out the excellent engine, the spit,
The plough and the flail, the mill and the hopper,
The hutch and the bolter, the furnace and copper:
The oven, the bavin, the malkin and peel,
The hearth and the range, the dog and the wheel.
He, he first invented both hogshead and tun,
The gimlet and vice, too, and taught 'em to run.
And since, with the funnel, an hippocras bag
He's made of himself, that now he cries swag.
Which shows, though the pleasure be but of four inches,
Yet he is a weezle, the gullet that pinches,
Of any delight; and not spares from the back
Whatever, to make of the belly a sack.
Hail, hail, plump paunch! O the founder of taste
For fresh meats or powdered or pickle or paste;
Devourer of broiled, baked, roasted or sod,
And emptier of cups, be they even or odd.
All which have now made thee so wide i' the waist
As scarce with no pudding thou art to be laced:
But eating and drinking until thou dost nod,
Thou breakst all thy girdles, and breakst forth a god.

hutch] vessel for kneading *bolter*] sieve *bavin*] bundle of firewood
malkin] kitchen mop *peel*] baker's shovel *gimlet*] hole for a tap *vice*] a tap
hippocras bag] wine strainer *cries swag*] flaunts his belly *weezle*] windpipe
sod] boiled *pudding*] (secondary meaning) binding to support a ship's mast

ANONYMOUS

20

HA ha! ha ha! This world doth pass
   Most merrily I'll be sworn,
For many an honest Indian ass
   Goes for a unicorn.
                    Fara diddle dyno,
                    This is idle fyno.

Tie hie! tie hie! O sweet delight!
   He tickles this age that can
Call Tullia's ape a marmasyte
   And Leda's goose a swan.
                    Fara diddle dyno,
                    This is idle fyno.

So so! so so! Fine English days!
   For false play is no reproach,
For he that doth the coachman praise
   May safely use the coach.
                    Fara diddle dyno,
                    This is idle fyno.

RICHARD, BISHOP CORBET

1582–1635

21

The Distracted Puritan

AM I mad, O noble Festus,
When zeal and godly knowledge
   Have put me in hope
   To deal with the Pope,
As well as the best in the college?
   Boldly I preach, hate a cross, hate a surplice,
      Mitres, copes, and rotchets;
   Come hear me pray nine times a day,
      And fill your heads with crochets.

rotchet] a type of surplice worn by a bishop

In the house of pure Emanuel
I had my education,
    Where my friends surmise
    I dazzl'd my eyes
With the light of revelation.
    Boldly I preach, &c.

They bound me like a bedlam,
They lash'd my four poor quarters;
    Whilst this I endure,
    Faith makes me sure
To be one of Fox's martyrs.
    Boldly I preach, &c.

These injuries I suffer
Through Anti-Christ's persuasion:
    Take off this chain,
    Neither Rome nor Spain
Can resist my strong invasion.
    Boldly I preach, &c.

Of the beast's ten horns (God bless us!)
I have knock'd off three already;
    If they let me alone
    I'll leave him none:
But they say I am too heady.
    Boldly I preach, &c.

When I sack'd the Seven-hill'd City,
I met the great red Dragon;
    I kept him aloof
    With the armour of proof,
Though here I have never a rag on.
    Boldly I preach, &c.

With a fiery sword and target,
There fought I with this monster:
    But the sons of pride
    My zeal deride,
And all my deeds misconster.
    Boldly I preach, &c.

Emanuel] Emanuel College, Cambridge, generally regarded as the most puritan college in the University

I unhors'd the Whore of Babel,
With the lance of Inspiration;
  I made her stink,
  And spill the drink
In the cup of abominations.
  Boldly I preach, &c.

I have seen two in a vision
With a flying book between them.
  I have been in despair
  Five times a year,
And been cur'd by reading Greenham.
  Boldly I preach, &c.

I observ'd in Perkins' Tables
The black lines of damnation;
  Those crooked veins
  So stuck in my brains,
That I fear'd my reprobation.
  Boldly I preach, &c.

In the holy tongue of Canaan
I plac'd my chiefest pleasure:
  Till I prick'd my foot
  With an Hebrew root,
That I bled beyond all measure.
  Boldly I preach, &c.

I appear'd before the archbishop,
And all the high commission;
  I gave him no grace,
  But told him to his face,
That he favour'd superstition.
  Boldly I preach, hate a cross, hate a surplice,
    Mitres, copes, and rotchets:
  Come hear me pray nine times a day,
    And fill your heads with crotchets.

Greenham] Richard Greenham, puritan divine *Perkins'*] William Perkins, puritan theologian who drew up 'A Table Deciding the Order and Cause of Salvation and Damnation'

ANONYMOUS

22

Fair and Scornful

FAIR and scornful, do thy worst.
I can laugh until I burst.
Thou canst not breed me so much care
As will blanch or change a hair.

I cannot whine, I cannot cry,
Nor put the finger in the eye,
But I can scowl and look askew
And I can scorn as well as you.

Fair and scornful, do thy worst.
I can laugh until thou burst.

ANONYMOUS

23

If All the World Were Paper

IF all the world were paper,
   And all the sea were ink,
If all the trees were bread and cheese,
   How should we do for drink?

If all the world were sand O,
   Oh then what should we lack O,
If as they say there were no clay,
   How should we take tobacco?

If all our vessels ran-a,
   If none but had a crack-a,
If Spanish apes ate all the grapes,
   How should we do for sack-a?

If all the world were men,
   And men lived all in trenches,
And there were none but we alone,
   How should we do for wenches?

If friars had no bald pates,
Nor nuns had no dark cloisters,
If all the seas were beans and peas,
How should we do for oysters?

If there had been no projects,
Nor none that did great wrongs,
If fiddlers shall turn players all,
How should we do for songs?

If all things were eternal,
And nothing their end bringing,
If this should be, then how should we
Here make an end of singing?

ANONYMOUS

24

Lawyers

LAWYERS themselves uphold the commonweal.
They punish those that do offend and steal.
They free with cunning art the innocent
From danger, loss and causeless punishment.
They can but will not keep the world in awe
With misexpounded and distended law.
They always have great store of charity
And love they want not, keeping amity.

SAMUEL BUTLER

1612–1680

25

from *Hudibras*

[*The Presbyterian Knight*]

FOR his religion it was fit
To match his learning and his wit:
'Twas Presbyterian true blue,
For he was of that stubborn crew
Of errant saints, whom all men grant
To be the true Church Militant:
Such as do build their faith upon
The holy text of pike and gun;
Decide all controversies by
Infallible artillery;
And prove their doctrine orthodox
By apostolic blows and knocks;
Call fire and sword and desolation,
A godly-thorough-Reformation,
Which always must be carry'd on,
And still be doing, never done:
As if religion were intended
For nothing else but to be mended.
A sect, whose chief devotion lies
In odd perverse antipathies;
In falling out with that or this,
And finding somewhat still amiss:
More peevish, cross, and splenetic,
Than dog distract, or monkey sick.
That with more care keep Holy-day
The wrong, than others the right way:
Compound for sins, they are inclin'd to;
By damning those they have no mind to;
Still so perverse and opposite,
As if they worshipp'd God for spite,
The self-same thing they will abhor
One way, and long another for.
Free-will they one way disavow,
Another, nothing else allow.
All piety consists therein
In them, in other men all sin.

CHARLES COTTON

1630–1687

26 from *Burlesque upon the Great Frost*

BUT, to leave fooling, I assure ye
There never was so cold a Fury
Of nipping Frost, and pinching weather
Since *Eve* and *Adam* met together.
Our *Peak*, that always has been famous
For cold wherewith to cramp and lame us,
Worse than it self, did now resemble a
Certain damn'd place call'd *Nova Zembla*,
And we who boast us humane Creatures,
Had happy been had we chang'd features,
Garments at least, though theirs be shabbed,
With those who that cold place inhabit,
The Bears and Foxes, who *sans* question
Than we by odds have warmer Vests on.
How cold that Country is, he knows most
Has there his Fingers and his Toes lost;
But here I know that every Member
Alike was handled by *December*,
Who blew his nose had clout or fist all
Instead of snivel fill'd with Crystal,
Who drew for Urinal ejection,
Was b'witch'd into an odd erection,
And these, *Priapus* like, stood strutting,
Fitter for Pedestal than rutting:
As men were fierce, or gentle handed,
Their Fists were clutch'd, or Palms expanded;
Limbs were extended, or contracted,
As use or humour most affected;
For, as men did to th' air expose 'em,
It catch'd and in that figure froze 'em;

. . . .

A Maid compell'd to be a gadder,
T' abate th' extension of her Bladder,
Which is an importuning matter,
Was so supported by her water,

To ease her knees with a third Pillar,
That as she sate the poor distiller
Look'd on the tripod, like the famous
Astrologer hight *Nostradamus*.

. . . .

Our Country Air was, in plain dealing,
Some weeks together so congealing,
That if, as men are rude in this age,
One spit had in another's visage,
The Constable by th' back had got him,
For he infallibly had shot him.
Nay, Friend with Friend, Brother with Brother,
Must needs have wounded one another
With kindest words, were they not wary
To make their greetings sideways carry;
For all the words that came from gullets,
If long were slugs, if short ones Bullets.
You might have read from mouths (*sans* Fable),
Your humble Servant, Sir, in Label;
Like those (yet theirs were warmer quarters),
We see in *Foxe's* Book of Martyrs.

ANONYMOUS

27

My Mistress

My mistress loves no woodcocks
Yet loves to pick the bones;
My mistress loves no jewels
Yet loves the precious stones;
My mistress loves no hunting
Yet loves to hear the horn;
My mistress loves no tables
Yet loves to see men lorn;
My mistress loves no wrestling
Yet loves to take a fall;
My mistress loves not some things,
And yet she loveth all;

My mistress loves a spender
Yet loves she not a waster;
My mistress loves no cuckold,
And yet she loves my master.

JOHN DRYDEN

1631–1700

28 from *Mac Flecknoe*

[*The Prince of Dulness Names His Heir*]

ALL human things are subject to decay,
And, when Fate summons, monarchs must obey:
This Flecknoe found, who, like Augustus, young
Was called to empire and had governed long:
In prose and verse was owned, without dispute
Through all the realms of Nonsense, absolute.
This aged prince now flourishing in peace,
And blest with issue of a large increase,
Worn out with business, did at length debate
To settle the succession of the state;
And pond'ring which of all his sons was fit
To reign, and wage immortal war with wit,
Cried, ''Tis resolved; for Nature pleads that he
Should only rule, who most resembles me;
Shadwell alone my perfect image bears,
Mature in dulness from his tender years;
Shadwell alone of all my sons is he
Who stands confirmed in full stupidity.
The rest to some faint meaning make pretence,
But Shadwell never deviates into sense.
Some beams of wit on other souls may fall,
Strike through and make a lucid interval;
But Shadwell's genuine night admits no ray,
His rising fogs prevail upon the day:
Besides, his goodly fabric fills the eye
And seems designed for thoughtless Majesty:
Thoughtless as Monarch Oaks that shade the plain,
And, spread in solemn state, supinely reign.'

29 Epilogue to *Tyrannic Love*

(Spoken by Mrs Ellen, when she had to be carried off dead by the bearers)

To the Bearer

HOLD, are you mad? You damned confounded dog,
I am to rise, and speak the Epilogue.

To the audience

I come, kind gentlemen, strange news to tell ye,
I am the ghost of poor departed Nelly.
Sweet ladies, be not frighted, I'll be civil,
I'm what I was, a little harmless devil.
For after death we sprites have just such natures
We had for all the world, when human creatures;
And therefore I that was an actress here,
Play all my tricks in Hell, a goblin there.
Gallants, look to it, you say there are no sprites;
But I'll come dance about your beds at nights.
And faith you'll be in a sweet kind of taking,
When I surprise you between sleep and waking.
To tell you true, I walk because I die
Out of my calling in a tragedy.
O poet, damned dull poet, who could prove
So senseless, to make Nelly die for love;
Nay, what's yet worse, to kill me in the prime
Of Easter term, in tart and cheesecake time!
I'll fit the fop; for I'll not one word say
To excuse his godly out-of-fashion play.
A play, which if you dare but twice sit out,
You'll all be slandered, and be thought devout.
But farewell, gentlemen, make haste to me,
I'm sure ere long to have your company.
As for my epitaph, when I am gone,
I'll trust no poet, but will write my own.

Here Nelly *lies, who, though she lived a slattern,*
Yet died a princess, acting in St Cathar'n.

Mrs Ellen] Nell Gwynn

THOMAS FLATMAN

1637–1688

30

An Appeal to Cats in the Business of Love

YE cats that at midnight spit at each other,
Who best feel the pangs of a passionate lover,
I appeal to your scratches and your tattered fur
If the business of love be no more than to purr?
Old Lady Grimalkin with her gooseberry eyes
Knew something when a kitten—for why? she was wise;
You find by experience the love-fit's soon o'er:
Puss-puss! lasts not long but turns to Cat-whore!
Men ride many miles,
Cats tread many tiles,
Both hazard their necks in the fray;
Only cats, when they fall
From a house or a wall,
Keep their feet, mount their tails, and away!

SIR CHARLES SEDLEY

1639?–1701

31

On a Cock at Rochester

THOU cursed cock, with thy perpetual noise,
Mayst thou be capon made, and lose thy voice,
Or on a dunghill mayst thou spend thy blood,
And vermin prey upon thy craven brood;
May rivals tread thy hens before thy face,
Then with redoubled courage give thee chase;
Mayst thou be punished for St Peter's crime,
And on Shrove Tuesday perish in thy prime;
May thy bruised carcass be some beggar's feast,
Thou first and worst disturber of man's rest.

JOHN WILMOT, EARL OF ROCHESTER

1647–1680

32

Upon Nothing

NOTHING! thou elder brother even to shade,
That hadst a being ere the world was made,
And (well fixed) art alone of ending not afraid.

Ere Time and Place were, Time and Place were not,
When primitive Nothing Something straight begot,
Then all proceeded from the great united—What?

Something, the general attribute of all,
Sever'd from thee, its sole original,
Into thy boundless self must undistinguish'd fall.

Yet Something did thy mighty power command,
And from thy fruitful emptiness's hand
Snatch'd men, beasts, birds, fire, air, and land.

Matter, the wicked'st offspring of thy race,
By Form assisted, flew from thy embrace;
And rebel Light obscured thy reverend dusky face.

With Form and Matter, Time and Place did join;
Body, thy foe, with thee did leagues combine,
To spoil thy peaceful realm, and ruin all thy line.

But turn-coat Time assists the foe in vain,
And, bribed by thee, assists thy short-liv'd reign,
And to thy hungry womb drives back thy slaves again.

Though mysteries are barr'd from laic eyes,
And the divine alone, with warrant, pries
Into thy bosom, where the truth in private lies;

Yet this of thee the wise may freely say,
Thou from the virtuous Nothing tak'st away,
And to be part with thee the wicked wisely pray

Great Negative! how vainly would the wise
Inquire, define, distinguish, teach, devise?
Didst thou not stand to point their dull philosophies.

Is, or *is not*, the two great ends of Fate,
And, true or false, the subject of debate,
That perfect or destroy the vast designs of Fate;

When they have rack'd the politician's breast,
Within thy bosom most securely rest,
And, when reduced to thee, are least unsafe and best.

But Nothing, why does Something still permit,
That sacred monarchs should at council sit,
With persons highly thought at best for nothing fit?

Whilst weighty Something modestly abstains
From princes' coffers, and from statesmen's brains,
And nothing there like stately Nothing reigns.

Nothing, who dwell'st with fools in grave disguise,
For whom they reverend shapes and forms devise,
Lawn sleeves, and furs, and gowns, when they like thee look wise.

French truth, Dutch prowess, British policy,
Hibernian learning, Scotch civility,
Spaniards' dispatch, Danes' wit, are mainly seen in thee.

TOM BROWN

1663–1704

33

Doctor Fell

I DO not love thee, Doctor Fell.
The reason why, I cannot tell;
But this I know, and know full well,
I do not love thee, Doctor Fell.

34

Oaths

OUR fathers took oaths as of old they took wives,
To have and to hold for the term of their lives,
But we take our oaths, as our whores, for our ease,
And a whore and a rogue may part when they please.

MATTHEW PRIOR

1664–1721

35

An Epitaph

INTERR'D beneath this marble stone
Lie Saunt'ring JACK, and Idle JOAN,
While rolling threescore years and one
Did round this globe their courses run;
If human things went ill or well;
If changing empires rose or fell;
The morning past, the evening came,
And found this couple still the same.
They walk'd and eat, good folks: What then?
Why then they walk'd and eat again:
They soundly slept the night away:
They did just nothing all the day:
And having buried children four,
Would not take pains to try for more.
Nor sister either had, nor brother:
They seem'd just tallied for each other.

Their moral and economy
Most perfectly they made agree:
Each virtue kept its proper bound,
Nor trespass'd on the other's ground.
Nor fame, nor censure they regarded:
They neither punish'd, nor rewarded.
He car'd not what the footmen did:
Her maids she neither prais'd, nor chid:
So ev'ry servant took his course;
And bad at first, they all grew worse.

Slothful disorder fill'd his stable;
And sluttish plenty deck'd her table.
Their beer was strong; their wine was port;
Their meal was large; their grace was short.
They gave the poor the remnant-meat,
Just when it grew not fit to eat.

They paid the church and parish rate;
And took, but read not the receipt:
For which they claim'd their Sunday's due,
Of slumb'ring in an upper pew.

No man's defects sought they to know;
So never made themselves a foe.
No man's good deeds did they commend;
So never rais'd themselves a friend.
Nor cherish'd they relations poor:
That might decrease their present store:
Nor barn nor house did they repair:
That might oblige their future heir.

They neither added, nor confounded:
They neither wanted, nor abounded.
Each Christmas they accompts did clear;
And wound their bottom round the year.
Nor tear, nor smile did they imploy
At news of public grief, or joy.
When bells were rung, and bonfires made,
If ask'd, they ne'er denied their aid:
Their jug was to the ringers carried.
Whoever either died, or married.
Their billet at the fire was found,
Whoever was depos'd, or crown'd.

Nor good, nor bad, nor fools, nor wise,
They would not learn, nor could advise:
Without love, hatred, joy, or fear,
They led—a kind of—as it were:
Nor wish'd, nor car'd, nor laugh'd, nor cried:
And so they liv'd; and so they died.

36

Human Life

WHAT trifling coil do we poor mortals keep;
Wake, eat, and drink, evacuate, and sleep.

37

A Dutch Proverb

FIRE, water, woman, are man's ruin,
Says wise Professor Vander Bruin.
By flames a house I hir'd was lost
Last year: and I must pay the cost.
This spring the rains o'erflow'd my ground:
And my best Flanders mare was drown'd.
A slave I am to Clara's eyes:
The gipsy knows her pow'r, and flies.
Fire, water, woman, are my ruin.
And great thy wisdom, Vander Bruin.

38

Written in an Ovid

OVID is the surest guide
You can name, to show the way
To any woman, maid or bride,
Who resolves to go astray.

39

Epitaph on Himself

NOBLES and heralds, by your leave,
Here lies what once was Matthew Prior;
The son of Adam and of Eve,
Can Bourbon or Nassau go higher?

NED WARD

1667–1731

40

A South Sea Ballad

In London stands a famous pile,
   And near that pile an Alley,
Where merry crowds for riches toil,
   And wisdom stoops to folly.
Here, sad and joyful, high and low,
   Court Fortune for her graces;
And as she smiles or frowns, they show
   Their gestures and grimaces.

Here, Stars and Garters do appear
   Among our lords the rabble;
To buy and sell, to see and hear
   The Jews and Gentiles squabble.
Here, crafty Courtiers are too wise
   For those who trust to fortune;
They see the cheat with clearer eyes,
   Who peep behind the curtain.

Long heads may thrive, by sober rules;
   Because they think, and drink not;
But headlongs are our thriving fools,
   Who only drink, and think not.
The lucky rogues like spaniel dogs,
   Leap into South Sea water;
And there they fish for golden frogs,
   Nor caring what comes after.

'Tis said that alchemists of old
   Could turn a brazen kettle,
Or leaden cistern into gold;
   That noble tempting metal.
But (if it here may be allowed,
   To bring in great with small things)
Our cunning South Sea like a god,
   Turns nothing into all things.

What need have we of Indian wealth,
  Or commerce with our neighbours;
Our Constitution is in health,
  And riches crown our labours.
Our South Sea ships have golden shrouds,
  They bring us wealth, 'tis granted:
But lodge their treasure in the clouds,
  To hide it till it's wanted.

O, Britain! bless thy present state!
  Thou only happy nation!
So oddly rich, so madly great,
  Since Bubbles came in fashion.
Successful rakes exert their pride,
  And count their airy millions;
Whilst homely drabs in coaches ride,
  Brought up to Town on pillions.

Few men who follow reason's rules,
  Grow fat with South Sea diet;
Young rattles and unthinking fools
  Are those that flourish by it.
Old musty jades, and pushing blades,
  Who've least consideration,
Grow rich apace; while wiser heads
  Are struck with admiration.

A race of men, who, t' other day,
  Lay crushed beneath disasters,
Are now, by Stock, brought into play,
  And made our lords and masters.
But should our South Sea Babel fall,
  What numbers would be frowning;
The losers then must ease their gall
  By hanging, or by drowning.

Bubbles] the South Sea Bubble, a speculative boom in the shares of the South Sea Company, which failed disastrously in 1720

Five hundred millions, notes and bonds,
  Our Stocks are worth in value:
But neither lie in goods, or lands,
  Or money, let me tell ye.
Yet though our foreign trade is lost,
  Of mighty wealth we vapour;
When all the riches that we boast
  Consist of scraps of paper.

JONATHAN SWIFT

1667–1745

41

Twelve Articles

1. LEST it may more quarrels breed
   I will never hear you read.
2. By disputing I will never
   To convince you, once endeavour.
3. When a paradox you stick to,
   I will never contradict you.
4. When I talk, and you are heedless,
   I will show no anger needless.
5. When your speeches are absurd,
   I will ne'er object one word.
6. When you furious argue wrong,
   I will grieve, and hold my tongue.
7. Not a jest, or humorous story,
   Will I ever tell before ye:
   To be chidden for explaining
   When you quite mistake the meaning.
8. Never more will I suppose
   You can taste my verse or prose:
9. You no more at me shall fret,
   While I teach, and you forget;
10. You shall never hear me thunder,
    When you blunder on, and blunder.
11. Show your poverty of spirit,
    And in dress place all your merit;
    Give yourself ten thousand airs
    That with me shall break no squares.

12. Never will I give advice
Till you please to ask me thrice;
Which, if you in scorn reject,
'Twill be just as I expect.

Thus we both shall have our ends,
And continue special friends.

42 *Phyllis*, or *The progress of Love*

DESPONDING Phyllis was endued
With every talent of a prude:
She trembled when a man drew near;
Salute her, and she turned her ear:
If o'er against her you were placed
She durst not look above your waist:
She'd rather take you to her bed,
Than let you see her dress her head;
In church you heard her, through the crowd
Repeat the absolution loud;
In church, secure behind her fan
She durst behold that monster, man:
There practised how to place her head.
And bit her lips to make them red;
Or on the mat devoutly kneeling
Would lift her eyes up to the ceiling,
And heave her bosom, unaware,
For neighbouring beaux to see it bare.

At length a lucky lover came,
And found admittance to the dame.
Suppose all parties now agreed,
The writings drawn, the lawyer fee'd,
The vicar and the ring bespoke:
Guess, how could such a match be broke?
See then what mortals place their bliss in!
Next morn betimes the bride was missing.
The mother screamed, the father chid;
Where can this idle wretch be hid?
No news of Phyl! The bridegroom came,
And thought his bride had skulked for shame,
Because her father used to say
The girl had such a bashful way.

Now John, the butler, must be sent
To learn the road that Phyllis went;
The groom was wished to saddle Crop;
For John must neither light nor stop;
But find her whereso'er she fled,
And bring her back, alive or dead.

See here again the devil to do;
For truly John was missing too.
The horse and pillion both were gone!
Phyllis, it seems, was fled with John.

Old Madam, who went up to find
What papers Phyl had left behind,
A letter on the toilet sees,
To my much honoured father,—these:
('Tis always done, romances tell us,
When daughters run away with fellows)
Filled with the choicest commonplaces,
By others used in the like cases;
'That, long ago a fortune-teller
Exactly said what now befell her;
And in a glass had made her see
A serving-man of low degree.
It was her fate, must be forgiven,
For marriages were made in heaven:
His pardon begged, but to be plain,
She'd do't if 'twere to do again.
Thank God, 'twas neither shame nor sin;
For John was come of honest kin.
Love never thinks of rich and poor,
She'd beg with John from door to door:
Forgive her, if it be a crime,
She'll never do't another time.
She ne'er before in all her life
Once disobeyed him, maid nor wife.
One argument she summed up all in,
The thing was done and past recalling:
And therefore hoped she should recover
His favour, when his passion's over.
She valued not what others thought her,
And was—his most obedient daughter.'

Fair maidens all attend the muse
Who now the wandering pair pursues.
Away they rode in homely sort,
Their journey long, their money short;
The loving couple well bemired;
The horse and both the riders tired:
Their victuals bad, their lodging worse;
Phyl cried, and John began to curse;
Phyl wished, that she had strained a limb,
When first she ventured out with him:
John wished, that he had broke a leg
When first for her he quitted Peg.

But what adventures more befell 'em,
The muse hath now no time to tell 'em.
How Johnny wheedled, threatened, fawned,
Till Phyllis all her trinkets pawned:
How oft she broke her marriage vows
In kindness to maintain her spouse,
Till swains unwholesome spoiled the trade;
For now the surgeon must be paid,
To whom those perquisites are gone,
In Christian justice due to John.

When food and raiment now grew scarce,
Fate put a period to the farce,
And with exact poetic justice;
For John is landlord, Phyllis hostess:
They keep, at Staines, the Old Blue Boar,
Are cat and dog, and rogue and whore.

43

The Description of an Irish Feast

(*Translated Almost Literally out of the Original Irish*)

O'ROURK's noble fare
  Will ne'er be forgot,
By those who were there,
  And those who were not.
His revels to keep,
  We sup and we dine,
On seven score sheep,
  Fat bullock and swine.

Usquebaugh to our feast
  In pails was brought up,
An hundred at least,
  And a madder our cup.
O there is the sport,
  We rise with the light,
In disorderly sort,
  From snoring all night.
O how was I tricked,
  My pipe it was broke,
My pocket was picked,
  I lost my new cloak.
I'm rifled, quoth Nell,
  Of mantle and kercher,
Why then fare them well,
  The de'il take the searcher.
Come, harper, strike up,
  But first by your favour,
Boy, give us a cup;
  Ay, this has some savour:
O'Rourk's jolly boys
  Ne'er dreamt of the matter,
Till roused by the noise,
  And musical clatter,
They bounce from their nest,
  No longer will tarry,
They rise ready dressed,
  Without one *Ave Mary*.
They dance in a round,
  Cutting capers and ramping,
A mercy the ground
  Did not burst with their stamping,
The floor is all wet
  With leaps and with jumps,
While the water and sweat,
  Splishsplash in their pumps.
Bless you late and early,
  Laughlin O' Enagin,
By my hand you dance rarely,
  Margery Grinagin.

Usquebaugh] whiskey *madder*] a wooden vessel

Bring straw for our bed,
   Shake it down to the feet,
Then over us spread,
   The winnowing sheet.
To show, I don't flinch,
   Fill the bowl up again,
Then give us a pinch
   Of your sneezing, a Yean.
Good Lord, what a sight,
   After all their good cheer,
For people to fight
   In the midst of their beer:
They rise from their feast,
   And hot are their brains,
A cubit at least
   The length of their skenes.
What stabs and what cuts,
   What clattering of sticks,
What strokes on the guts,
   What bastings and kicks!
With cudgels of oak,
   Well hardened in flame,
An hundred heads broke,
   An hundred struck lame.
You churl, I'll maintain
   My father built Lusk,
The castle of Slane,
   And Carrickdrumrusk:
The Earl of Kildare,
   And Moynalta, his brother,
As great as they are,
   I was nursed by their mother.
Ask that of old Madam,
   She'll tell you who's who,
As far up as Adam,
   She knows it is true,
Come down with that beam,
   If cudgels are scarce,
A blow on the wame,
   Or a kick on the arse.

Yean] an Irish name for a woman   *skenes*] short swords   *wame*] belly

44

On a Curate's Complaint of Hard Duty

I MARCH'D three miles through scorching sand,
With zeal in heart, and notes in hand;
I rode four more to Great St Mary,
Using four legs, when two were weary:
To three fair virgins I did tie men,
In the close bands of pleasing Hymen;
I dipp'd two babes in holy water,
And purified their mother after.
Within an hour and eke a half,
I preach'd three congregations deaf;
Where, thundering out, with lungs long-winded,
I chopp'd so fast, that few there minded.
My emblem, the laborious sun,
Saw all these mighty labours done
Before one race of his was run.
All this perform'd by Robert Hewit:
What mortal else could e'er go through it!

WILLIAM CONGREVE

1670–1729

45

Semele to Jupiter

WITH my frailty, don't upbraid me,
I am woman as you made me;
Causeless doubting, or despairing,
Rashly trusting, idly fearing:
        If obtaining,
        Still complaining;
        If consenting,
        Still repenting;
        Most complying,
        Then denying:
And to be followed, only flying.

With my frailty, don't upbraid me:
I am woman as you made me.

ANONYMOUS

46

Brian O Linn

BRIAN O LINN had no breeches to wear
He got an old sheepskin to make him a pair
With the fleshy side out and the woolly side in,
'They'll be pleasant and cool,' says Brian O Linn.

Brian O Linn had no shirt to his back,
He went to a neighbour's, and borrowed a sack,
Then he puckered the meal bag in under his chin—
'Sure they'll take them for ruffles,' says Brian O Linn.

Brian O Linn was hard up for a coat,
So he borrowed the skin of a neighbouring goat,
With the horns sticking out from his oxsters, and then,
'Sure they'll take them for pistols,' says Brian O Linn.

Brian O Linn had no hat to put on,
So he got an old beaver to make him a one,
There was none of the crown left and less of the brim,
'Sure there's fine ventilation,' says Brian O Linn.

Brian O Linn had no brogues for his toes,
He hopped in two crab-shells to serve him for those.
Then he split up two oysters that match'd like a twin,
'Sure they'll shine out like buckles,' says Brian O Linn.

Brian O Linn had no watch to put on,
So he scooped out a turnip to make him a one.
Then he placed a young cricket in—under the skin—
'Sure they'll think it is ticking,' says Brian O Linn.

Brian O Linn to his house had no door,
He'd the sky for a roof, and the bog for a floor;
He'd a way to jump out, and a way to swim in,
''Tis a fine habitation,' says Brian O Linn.

Brian O Linn went a-courting one night,
He set both the mother and daughter to fight;
To fight for his hand they both stripped to the skin,
'Sure! I'll marry you both,' says Brian O Linn.

Brian OLinn, his wife and wife's mother,
They all lay down in the bed together,
The sheets they were old and the blankets were thin,
'Lie close to the wall,' says Brian OLinn.

Brian OLinn, his wife and wife's mother,
Were all going home o'er the bridge together,
The bridge it broke down, and they all tumbled in,
'We'll go home by the water,' says Brian OLinn.

GEORGE FARQUHAR

1678–1707

47

Trifles

A TRIFLING song you shall hear;
  Begun with a trifle and ended.
All trifling people draw near,
  And I shall be nobly attended.

Were it not for trifles a few,
  That lately have come into play;
The men would want something to do,
  And the women want something to say.

What makes men trifle in dressing?
  Because the ladies, they know,
Admire, by often possessing,
  That eminent trifle, a Beau.

What mortal man would be able
  At White's half an hour to sit,
Or who could bear a tea-table,
  Without talking of trifles for wit?

The Court is from trifles secure;
  Gold Keys are no trifles, we see;
White rods are no trifles, I'm sure,
  Whatever their bearers may be.

But if you will go to the place
   Where trifles abundantly breed,
The Levée will show you his Grace
   Makes promises trifles indeed.

A coach with six footmen behind,
   I count neither trifle, nor sin;
But, ye gods! how oft do we find
   A scandalous trifle within.

A flask of champagne, people think it
   A trifle, or something as bad;
But if you'll contrive how to drink it,
   You'll find it no trifle, egad!

A parson's a trifle at sea,
   A widow's a trifle in sorrow;
A peace is a trifle to-day;
   Who knows what may happen to-morrow?

A black coat, a trifle may cloak;
   Or to hide it, the red may endeavour;
But if once the army is broke,
   We shall have more trifles than ever.

The stage is a trifle, they say;
   The reason, pray carry along;
Because at ev'ry new play,
   The house they with trifles so throng.

But with people's malice to trifle,
   And to set us all on a foot;
The author of this is a trifle,
   And his Song is a trifle to boot.

JOHN GAY

1685–1732

48

Molly Mog

I

SAYS my Uncle, I pray you discover
  What hath been the Cause of your Woes,
Why you pine, and you whine, like a Lover?
  I have seen *Molly Mog* of the *Rose*.

II

O Nephew! your Grief is but Folly,
  In Town you may find better Prog;
Half a Crown there will get you a *Molly*,
  A *Molly* much better than *Mog*.

III

I know that by Wits 'tis recited,
  That Women at best are a Clog;
But I am not so easily frighted,
  From loving of sweet *Molly Mog*.

IV

The School-Boy's desire is a Play-Day,
  The School-Master's joy is to flog;
The Milk-Maid's delight is on *May-Day*,
  But mine is on sweet *Molly Mog*.

V

Will-a-wisp leads the Trav'ler a gadding
  Thro' Ditch, and thro' Quagmire and Bog;
But no Light can set me a madding,
  Like the Eyes of my sweet *Molly Mog*.

VI

For Guineas in other Men's Breeches
  Your Gamesters will palm and will cog;
But I envy them none of their Riches,
  So I may win sweet *Molly Mog*.

Prog] provisions

VII

The Heart, when half-wounded, is changing,
It here and there leaps like a Frog;
But my Heart can never be ranging,
'Tis so fix'd upon sweet *Molly Mog*.

VIII

Who follows all Women of Pleasure
In Love, has a Taste like a Hog;
For no Girl can give better Measure
Of Joys, than my sweet Molly Mog.

IX

I feel I'm in Love to Distraction,
My Senses all lost in a Fog;
And nothing can give Satisfaction
But thinking of sweet *Molly Mog*.

X

A Letter when I am inditing,
Comes *Cupid* and gives me a Jog,
And I fill all the Paper with writing
Of nothing but sweet *Molly Mog*.

XI

If I would not give up the three *Graces*
I wish I were hang'd like a Dog,
And at Court all the Drawing-Room Faces,
For a Glance of my sweet *Molly Mog*.

XII

Those Faces want Nature and Spirit,
And seem as cut out of a Log;
Juno, Venus, and *Pallas*'s Merit
Unite in my sweet *Molly Mog*.

XIII

Those who toast all the Family Royal
In Bumpers of Hogan and Nog,
Can't have Hearts more true, nor more loyal,
Than mine is for sweet Molly Mog.

XIV

Were *Virgil* alive with his *Phillis*,
And writing another Eclogue;
Both his *Phillis* and fair *Amaryllis*
He'd give up for sweet *Molly Mog*.

XV

When she smiles on each Guest, like her Liquor,
Then Jealousy sets me agog.
To be sure she's a Bit for the *Vicar*,
And so I shall lose *Molly Mog*.

49 *To a Young Lady with Some Lampreys*

WITH lovers 'twas of old the fashion
By presents to convey their passion:
No matter what the gift they sent,
The lady saw that love was meant.
Fair Atalanta, as a favour,
Took the boar's head her hero gave her;
Nor could the bristly thing affront her,
'Twas a fit present from a hunter.
When squires send woodcocks to the dame,
It serves to show their absent flame:
Some by a snip of woven hair
In posied lockets bribe the fair;
How many mercenary matches
Have sprung from di'mond-rings and watches!
But hold—a ring, a watch, a locket,
Would drain at once a poet's pocket;
He should send songs that cost him nought,
Nor ev'n be prodigal of thought.
Why then send lampreys? fie, for shame!
'Twill set a virgin's blood on flame.
This to fifteen a proper gift!
It might lend sixty-five a lift.
I know your maiden aunt will scold,
And think my present somewhat bold.
I see her lift her hands and eyes:
'What, eat it, niece? eat Spanish flies!
Lamprey's a most immodest diet:
You'll neither wake nor sleep in quiet.

Should I tonight eat sago-cream,
'Twould make me blush to tell my dream;
If I eat lobster, 'tis so warming
That ev'ry man I see looks charming;
Wherefore had not the filthy fellow
Laid Rochester upon your pillow?
I vow and swear, I think the present
Had been as modest and as decent.
'Who has her virtue in her power?
Each day has its unguarded hour;
Always in danger of undoing,
A prawn, a shrimp may prove our ruin!
'The shepherdess, who lives on salad,
To cool her youth controls her palate;
Should Dian's maids turn liqu'rish livers,
And of huge lampreys rob the rivers,
Then all beside each glade and visto,
You'd see nymphs lying like Calisto.
'The man who meant to heat your blood,
Needs not himself such vicious food—'
In this, I own, your aunt is clear,
I sent you what I well might spare:
For when I see you (without joking),
Your eyes, lips, breasts are so provoking,
They set my heart more cock-a-hoop
Than could whole seas of craw-fish soup.

50 *The Two Monkeys*

THE learned, full of inward pride,
The fops of outward show deride;
The fop, with learning at defiance,
Scoffs at the pedant and the science:
The *Don*, a formal, solemn strutter,
Despises *Monsieur*'s airs and flutter:
While *Monsieur* mocks the formal fool,
Who looks, and speaks, and walks by rule.
Britain, a medly of the twain,
As pert as *France*, as grave as *Spain*.
In fancy wiser than the rest,
Laughs at them both, of both the jest.
Is not the poet's chiming close
Censur'd, by all the sons of prose?
While bards of quick imagination
Despise the sleepy prose narration.

Men laugh at apes, they men contemn;
For what are we, but apes to them?

Two Monkeys went to *Southwark* fair
No criticks had a sourer air.
They forc'd their way through draggled folks,
Who gap'd to catch *Jack-Pudding*'s jokes.
Then took their tickets for the show,
And got by chance the foremost row.
To see their grave observing face
Provok'd a laugh thro' all the place.
Brothers, says Pug, and turn'd his head,
The rabble's monstrously ill-bred.
Now through the booth loud hisses ran;
Nor ended 'till the Show began.
The tumbler whirles the flip-flap round,
With sommersets he shakes the ground;
The cord beneath the dancer springs;
Aloft in air the vaulter swings,
Distorted now, now prone depends,
Now through his twisted arms ascends;
The croud, in wonder and delight,
With clapping hands applaud the sight.
With smiles, quoth Pug; If pranks like these
The giant apes of reason please,
How would they wonder at our arts!
They must adore us for our parts.
High on the twig I've seen you cling,
Play, twist and turn in airy ring;
How can those clumsy things, like me,
Fly with a bound from tree to tree?
But yet, by this applause, we find
These emulators of our kind
Discern our worth, our parts regard,
Who our mean mimicks thus reward.
Brother, the grinning mate replies,
In this I grant that man is wise,
While good example they pursue,
We must allow some praise is due;
But when they strain beyond their guide,
I laugh to scorn the mimic pride.
For how fantastick is the sight,
To meet men always bolt upright,
Because we sometimes walk on two!
I hate the imitating crew.

ANONYMOUS

51

The Vicar of Bray

In good King Charles's golden days,
  When loyalty no harm meant;
A furious high-church man I was,
  And so I gain'd preferment.
Unto my flock I daily preach'd,
  Kings are by God appointed,
And damn'd are those who dare resist,
  Or touch the Lord's anointed.
    And this is law, I will maintain
      Unto my dying day, Sir,
    That whatsoever King shall reign,
      I will be Vicar of Bray, Sir!

When Royal James possessed the crown,
  And popery grew in fashion;
The penal law I hooted down,
  And read the declaration:
The Church of Rome, I found would fit,
  Full well my constitution,
And I had been a Jesuit,
  But for the Revolution.
    And this is law, &c.

When William our deliverer came,
  To heal the nation's grievance,
I turned the cat in pan again,
  And swore to him allegiance:
Old principles I did revoke,
  Set conscience at a distance,
Passive obedience is a joke,
  A jest is non-resistance.
    And this is law, &c.

When glorious Anne became our Queen,
  The Church of England's glory,
Another face of things was seen,
  And I became a Tory:

Occasional conformists base,
  I damn'd, and moderation,
And thought the church in danger was,
  From such prevarication.
    And this is law, &c.

When George in pudding time came o'er,
  And moderate men looked big, Sir,
My principles I chang'd once more,
  And so became a Whig, Sir:
And thus preferment I procur'd,
  From our Faith's Great Defender,
And almost every day abjur'd
  The Pope, and the Pretender.
    And this is law, &c.

The illustrious House of Hanover,
  And Protestant succession,
To these I lustily will swear,
  Whilst they can keep possession:
For in my faith, and loyalty,
  I never once will falter,
But George, my lawful King shall be,
  Except the times should alter.
    And this is law, &c.

HENRY CAREY

1687?–1743

52 *A Lilliputian Ode on their Majesties' Accession*

SMILE, smile,
Blest isle!
Grief past,
At last,
Halcyon
Comes on.
New King,
Bells ring;

New Queen,
Blest scene!
Britain
Again
Revives
And thrives.
Fear flies,
Stocks rise;
Wealth flows,
Art grows.
Strange pack
Sent back;
Own folks
Crack jokes.
Those out
May pout;
Those in
Will grin.

Great, small,
Pleased all.

God send
No end
To line
Divine
Of George and Caroline!

ALEXANDER POPE

1688–1744

53 *Imitation of Chaucer*

WOMEN ben full of Ragerie,
Yet swinken not sans secresie.
Thilke Moral shall ye understand,
From Schoole-boy's Tale of fayre Irelond:
Which to the Fennes hath him betake,
To filch the gray Ducke fro the Lake.

Right then, there passen by the Way
His Aunt, and eke her Daughters tway.
Ducke in his Trowses hath he hent,
Not to be spied of Ladies gent.
'But ho! our Nephew,' (crieth one)
'Ho,' quoth another, 'Cozen John';
And stoppen, and laugh, and callen out,—
This sely Clerk full low doth lout:
They asken that, and talken this,
'Lo here is Coz, and here is Miss.'
But, as he glozeth with Speeches soote,
The Ducke sore tickleth his Erse-root:
Fore-piece and buttons all-to-brest,
Forth thrust a white neck, and red crest.
'Te-he,' cry'd Ladies; Clerke nought spake:
Miss star'd; and gray Ducke crieth Quake.
'O Moder, Moder' (quoth the daughter)
'Be thilke same thing Maids longen a'ter?
'Bette is to pyne on coals and chalke,
'Then trust on Mon, whose yerde can talke.'

54 from *An Essay on Criticism*

WHERE'ER you find 'the cooling western breeze',
In the next line, it 'whispers through the trees';
If crystal streams 'with pleasing murmurs creep',
The reader's threatened (not in vain) with 'sleep'.
Then, at the last and only couplet fraught
With some unmeaning thing they call a thought,
A needless Alexandrine ends the song,
That, like a wounded snake, drags its slow length along.

55 *A Farewell to London, in the Year 1715*

DEAR, damn'd distracting town, farewell!
Thy fools no more I'll tease:
This year in peace, ye critics, dwell,
Ye harlots, sleep at ease!

Why should I stay? Both parties rage;
My vixen mistress squalls;
The wits in envious feuds engage;
And Homer (damn him!) calls.

Why make I friendships with the great,
  When I no favour seek?
Or follow girls seven hours in eight?
  I need but once a week.

Luxurious lobster-nights, farewell,
  For sober studious days!
And Burlington's delicious meal,
  For salads, tarts, and pease!

Adieu to all but Gay alone,
  Whose soul, sincere and free,
Loves all mankind, but flatters none,
  And so may starve with me.

56 from *Epistle to Arbuthnot*

SHUT, shut the door, good *John*! fatigu'd, I said,
Tye up the knocker! say I'm sick, I'm dead.
The Dog-star rages! nay, 'tis past a doubt,
All Bedlam, or Parnassus, is let out:
Fire in each eye, and papers in each hand,
They rave, recite, and madden round the land.
  What walls can guard me, or what shades can hide?
They pierce my thickets, thro' my Grot they glide,
By land, by water, they renew the charge,
They stop the chariot, and they board the Barge.
No place is sacred, not the Church is free,
Ev'n Sunday shines no Sabbath-day to me:
Then from the Mint walks forth the man of rhyme,
Happy! to catch me just at Dinner-time.
  Is there a Parson much be-mus'd in beer,
A maudlin Poetess, a rhyming Peer,
A Clerk, foredoom'd his father's soul to cross,
Who pens a Stanza, when he should *engross*?
Is there, who, lock'd from ink and paper, scrawls
With desp'rate charcoal round his darken'd walls?
All fly to *Twit'nam*, and in humble strain
Apply to me, to keep them mad or vain.
Arthur, whose giddy Son neglects the Laws,
Imputes to me and my damn'd works the cause:
Poor *Cornus* sees his frantic wife elope,
And curses Wit, and Poetry, and *Pope*.

good John] Pope's manservant

57

Visiting Dr Swift

How foolish men on expeditions go!
Unweeting wantons of their wetting woe!
For drizzling damps descend adown the plain
And seem a thicker dew, or thinner rain;
Yet dew or rain may wet us to the shift,
We'll not be slow to visit Dr Swift.

JOHN BYROM

1692–1763

58

from *Careless Content*

I AM content, I do not care,
  Wag as it will the world for me;
When fuss and fret was all my fare,
  It got no ground, as I could see:
So when away my caring went,
I counted cost, and was content.

With more of thanks, and less of thought,
  I strive to make my matters meet;
To seek what ancient sages sought,
  Physic and food, in sour and sweet:
To take what passes in good part,
And keep the hiccups from the heart.

With good and gentle-humoured hearts,
  I choose to chat where'er I come,
Whate'er the subject be that starts;
  But if I get among the glum,
I hold my tongue to tell the troth,
And keep my breath to cool my broth.

For chance or change, of peace or pain,
  For Fortune's favour or her frown;
For lack or glut, for loss or gain,
  I never dodge, nor up nor down:
But swing what way the ship shall swim,
Or tack about with equal trim.

I suit not where I shall not speed,
  Nor trace the turn of ev'ry tide;
If simple sense will not succeed,
  I make no bustling, but abide:
For shining wealth or scaring woe,
I force no friend, I fear no foe.

Of *Ups* and *Downs*, of *Ins* and *Outs*,
  Of *they're i' th' wrong*, and *we're i' th' right*,
I shun the rancours and the routs,
  And wishing well to every wight,
Whatever turn the matter takes,
I deem it all but ducks and drakes.

With whom I feast I do not fawn,
  Nor, if the folks should flout me, faint;
If wonted welcome be withdrawn,
  I cook no kind of a complaint:
With none disposed to disagree,
But like them best, who best like me.

. . . . . .

Now taste and try this temper, sirs,
  Mood it and brood it in your breast;
Or if ye ween, for worldly stirs,
  That man does right to mar his rest,
Let me be deft and debonair,
I am content, I do not care.

MATTHEW GREEN

1696–1737

59

from *The Spleen*

SOMETIMES I dress, with women sit,
And chat away the gloomy fit;
Quit the stiff garb of serious sense,
And wear a gay impertinence,
Nor think nor speak with any pains,
But lay on fancy's neck the reins;
Talk of unusual swell of waist
In maid of honour loosely laced,
And beauty borrowing Spanish red,
And loving pair with separate bed,
And jewels pawned for loss of game,
And then redeemed by loss of fame;
Of Kitty (aunt left in the lurch
By grave pretence to go to church)
Perceived in hack with lover fine,
Like Will and Mary on the coin:
And thus in modish manner we,
In aid of sugar, sweeten tea.

GEORGE FAREWELL

fl. 1730

60

Privy-Love for my Landlady

HERE costive many minutes did I strain,
Still squeezing, sweating, swearing, all in vain;
When lo! who should pop by but mother Masters,
At whose bewitching look soon stubborn arse stirs.
No more my wanton wit shall whip thy wife,
Dear, doting Dick, for O! she saved my life.

61

Quaere

WHETHER at doomsday (tell, ye reverend wise)
My friend Priapus with myself shall rise?

SIR CHARLES HANBURY WILLIAMS

1708–1759

62

A Song upon Miss Harriet Hanbury, Addressed to the Revd Mr Birt

DEAR Doctor of St Mary's,
In the hundred of 'Bergavenny,
I've seen such a lass,
With a shape and a face,
As never was match'd by any.

Such wit, such bloom, and such beauty,
Has this girl of Ponty-Pool, Sir,
With eyes that would make
The toughest heart ache,
And the wisest man a fool, Sir.

At our fair t' other day she appear'd, Sir,
And the Welshmen all flock'd and view'd her;
And all of them said,
She was fit t' have been made
A wife for Owen Tudor.

They would ne'er have been tired of gazing,
And so much her charms did please, Sir,
That all of them sat
Till their ale grew flat,
And cold was their toasted cheese, Sir.

How happy the lord of the manor,
That shall be of her possest, Sir;
For all must agree,
Who my Harriet shall see,
She's a Harriet of the best, Sir.

Then pray make a ballad about her;
  We know you have wit if you'd show it,
    Then don't be ashamed,
    You can never be blamed,—
  For a prophet is often a poet!

But why don't you make one yourself, then?
  I suppose I by you shall be told, Sir,
    This beautiful piece
    Of Eve's flesh is my niece—
  And, besides, she's but five years old, Sir!

But tho', my dear friend, she's no older,
  In her face it may plainly be seen, Sir,
    That this angel at five,
    Will, if she's alive,
  Be a goddess at fifteen, Sir.

SAMUEL JOHNSON

1709–1784

63 from Prologue to Garrick's *Lethe*

PRODIGIOUS Madness of the writing Race!
Ardent of Fame, yet fearless of Disgrace.
  Without a boding Tear, or anxious Sigh,
The Bard obdurate sees his Brother die.
Deaf to the Critick, Sullen to the Friend,
Not One takes Warning, by another's End.

64 *Parody of Bishop Percy*

I THEREFORE pray thee, Renny dear,
  That thou wilt give to me,
With cream and sugar soften'd well,
  Another dish of tea.

Nor fear that I, my gentle maid,
  Shall long detain the cup,
When once unto the bottom I
  Have drank the liquor up.

Yet hear, alas! this mournful truth,
  Nor hear it with a frown;—
Thou canst not make the tea so fast
  As I can gulp it down.

65 *A Short Song of Congratulation*

LONG-EXPECTED one and twenty
  Ling'ring year at last is flown;
Pomp and pleasure, pride and plenty,
  Great Sir John, are all your own.

Loosened from the minor's tether,
  Free to mortgage or to sell,
Wild as wind and light as feather,
  Bid the slaves of thrift farewell.

Call the Bettys, Kates and Jennys,
  Ev'ry name that laughs at care;
Lavish of your grandsire's guineas,
  Show the spirit of an heir.

All that prey on vice and folly
  Joy to see their quarry fly:
Here the gamester light and jolly,
  There the lender grave and sly.

Wealth, Sir John, was made to wander,
  Let it wander as it will;
See the jockey, see the pander,
  Bid them come, and take their fill.

When the bonny blade carouses,
  Pockets full, and spirits high,
What are acres? What are houses?
  Only dirt, or wet or dry.

If the guardian or the mother
  Tell the woes of wilful waste,
Scorn their counsel and their pother,
  You can hang or drown at last.

JOHN BANKS

1709–1751

66 *A Description of London*

HOUSES, churches, mixed together,
Streets unpleasant in all weather;
Prisons, palaces contiguous,
Gates, a bridge, the Thames irriguous.

Gaudy things enough to tempt ye,
Showy outsides, insides empty;
Bubbles, trades, mechanic arts,
Coaches, wheelbarrows and carts.

Warrants, bailiffs, bills unpaid,
Lords of laundresses afraid;
Rogues that nightly rob and shoot men,
Hangmen, aldermen and footmen.

Lawyers, poets, priests, physicians,
Noble, simple, all conditions:
Worth beneath a threadbare cover,
Villainy bedaubed all over.

Women black, red, fair and grey,
Prudes and such as never pray,
Handsome, ugly, noisy, still,
Some that will not, some that will.

Many a beau without a shilling,
Many a widow not unwilling;
Many a bargain, if you strike it:
This is London! How d'ye like it?

THOMAS GRAY

1716–1771

67 *Ode on the Death of a Favourite Cat, Drowned in a Tub of Gold Fishes*

'TWAS on a lofty vase's side,
Where China's gayest art had dyed
 The azure flowers, that blow;
Demurest of the tabby kind,
The pensive Selima reclined,
 Gazed on the lake below.

Her conscious tail her joy declared;
The fair round face, the snowy beard,
 The velvet of her paws,
Her coat that with the tortoise vies,
Her ears of jet and emerald eyes,
 She saw; and purred applause.

Still had she gazed; but 'midst the tide
Two angel forms were seen to glide,
 The genii of the stream:
Their scaly armour's Tyrian hue
Through richest purple to the view
 Betrayed a golden gleam.

The hapless nymph with wonder saw:
A whisker first and then a claw,
 With many an ardent wish,
She stretched in vain to reach the prize.
What female heart can gold despise?
 What cat's averse to fish?

Presumptuous maid! with looks intent
Again she stretched, again she bent,
 Nor knew the gulf between.
(Malignant Fate sat by and smiled)
The slipp'ry verge her feet beguiled,
 She tumbled headlong in.

Eight times emerging from the flood
She mewed to every watry god,
    Some speedy aid to send.
No dolphin came, no Nereid stirred:
Nor cruel Tom nor Susan heard.
    A fav'rite has no friend!

From hence, ye beauties, undeceived,
Know, one false step is ne'er retrieved,
    And be with caution bold.
Not all that tempts your wand'ring eyes
And heedless hearts is lawful prize;
    Nor all that glisters gold.

JAMES CAWTHORN

1719–1761

68 from *Wit and Learning*

  Wit was a strange unlucky child,
Exceeding sly, and very wild;
Too volatile for truth or law,
He minded but his top or taw;
And, ere he reached the age of six,
Had played a thousand waggish tricks—
He drilled a hole in Vulcan's kettles,
He strewed Minerva's bed with nettles,
Climbed up the solar car to ride in't,
Broke off a prong from Neptune's trident,
Stole Amphitrite's favourite sea-knot,
And urined in Astrea's teapot.

JOHN CUNNINGHAM

1729–1773

69 *On Alderman W——: The History of his Life*

THAT he was born it cannot be denied,
He ate, drank, slept, talk'd politics, and died.

OLIVER GOLDSMITH

1730?–1774

70 *The Double Transformation*

SECLUDED from domestic strife,
Jack Book-worm led a college life;
A fellowship at twenty-five
Made him the happiest man alive;
He drank his glass and cracked his joke,
And freshmen wondered as he spoke.
  Such pleasures, unallayed with care,
Could any accident impair?
Could Cupid's shaft at length transfix
Our swain, arrived at thirty-six?
O had the archer ne'er come down
To ravage in a country town!
Or Flavia been content to stop
At triumphs in a Fleet-Street shop.
O had her eyes forgot to blaze!
Or Jack had wanted eyes to gaze.
O—but let exclamation cease,
Her presence banished all his peace.
So with decorum all things carried;
Miss frowned, and blushed, and then was—married.
  Need we expose to vulgar sight
The raptures of the bridal night?

Need we intrude on hallowed ground,
Or draw the curtains closed around?
Let it suffice that each had charms;
He clasped a goddess in his arms;
And, though she felt his usage rough,
Yet in a man 'twas well enough.
  The honey-moon like lightning flew,
The second brought its transports too.
A third, a fourth, were not amiss,
The fifth was friendship mixed with bliss:
But, when a twelvemonth passed away,
Jack found his goddess made of clay;
Found half the charms that decked her face
Arose from powder, shreds or lace;
But still the worst remained behind,
That very face had robbed her mind.
  Skilled in no other arts was she,
But dressing, patching, repartee;
And, just as humour rose or fell,
By turns a slattern or a belle:
'Tis true she dressed with modern grace,
Half naked at a ball or race;
But when at home, at board or bed,
Five greasy nightcaps wrapped her head.
Could so much beauty condescend
To be a dull domestic friend?
Could any curtain-lectures bring
To decency so fine a thing?
In short, by night 'twas fits or fretting;
By day 'twas gadding or coquetting.
Fond to be seen, she kept a bevy
Of powdered coxcombs at her levy;
The squire and captain took their stations,
And twenty other near relations;
Jack sucked his pipe and often broke
A sigh in suffocating smoke;
While all their hours were passed between
Insulting repartee or spleen.
  Thus as her faults each day were known,
He thinks her features coarser grown;
He fancies every vice she shows
Or thins her lip or points her nose:
Whenever rage or envy rise,
How wide her mouth, how wild her eyes!

He knows not know, but so it is,
Her face is grown a knowing phiz;
And, though her fops are wondrous civil,
He thinks her ugly as the devil.
 Now, to perplex the ravelled noose,
As each a different way pursues,
While sullen or loquacious strife
Promised to hold them on for life,
That dire disease, whose ruthless power
Withers the beauty's transient flower,
Lo! the small-pox with horrid glare
Levelled its terrors at the fair;
And, rifling every youthful grace,
Left but the remnant of a face.
 The glass, grown hateful to her sight,
Reflected now a perfect fright:
Each former art she vainly tries
To bring back lustre to her eyes.
In vain she tries her paste and creams,
To smooth her skin or hide its seams;
Her country beaux and city cousins,
Lovers no more, flew off by dozens:
The squire himself was seen to yield,
And even the captain quit the field.
 Poor Madam, now condemned to hack
The rest of life with anxious Jack,
Perceiving others fairly flown,
Attempted pleasing him alone.
Jack soon was dazzled to behold
Her present face surpass the old;
With modesty her cheeks are dyed,
Humility displaces pride;
For tawdry finery is seen
A person ever neatly clean:
No more presuming on her sway,
She learns good-nature every day;
Serenely gay and strict in duty,
Jack finds his wife a perfect beauty.

71 from *Retaliation*

[*Sir Joshua Reynolds*]

HERE Reynolds is laid and, to tell you my mind,
He has not left a better or wiser behind:
His pencil was striking, resistless and grand;
His manners were gentle, complying and bland;
Still born to improve us in every part,
His pencil our faces, his manners our heart;
To coxcombs averse, yet most civilly steering,
When they judged without skill he was still hard of hearing;
When they talked of their Raphaels, Correggios and stuff,
He shifted his trumpet and only took snuff.

ISAAC BICKERSTAFFE

1733–1808?

72 from *The Recruiting Serjeant: A Musical Entertainment*

[*Air*]

WHAT a charming thing's a battle!
Trumpets sounding, drums a-beating;
Crack, crick, crack, the cannons rattle,
Every heart with joy elating.
With what pleasure are we spying,
From the front and from the rear,
Round us in the smoky air,
Heads and limbs and bullets flying!
Then the groans of soldiers dying,
Just like sparrows as it were:
At each pop,
Hundreds drop,
While the muskets prittle prattle.
Killed and wounded
Lie confounded:
What a charming thing's a battle!
But the pleasant joke of all
Is when to close attack we fall,

Like mad bulls each other butting,
Shooting, stabbing, maiming, cutting;
Horse and foot
All go to 't,
Kill's the word, both men and cattle,
Then to plunder:
Blood and thunder,
What a charming thing's a battle!

73

An Expostulation

WHEN late I attempted your pity to move,
Why seem'd you so deaf to my pray'rs?
Perhaps it was right to dissemble your love—
But—Why did you kick me downstairs?

JOHN WOLCOT ('PETER PINDAR')

1738–1819

74

from *Bozzy and Piozzi, or The British Biographers*

A Town Eclogue

The Argument

On the Death of Dr Johnson, a Number of People, ambitious of being distinguished from the mute Part of their Species, set about relating and printing Stories and Bon-mots of that celebrated Moralist. Among the most zealous, though not the most enlightened, appeared Mr Boswell and Madame Piozzi, the Hero and Heroine of our Eclogue. They are supposed to have in Contemplation the Life of Johnson; and, to prove their biographical Abilities, appeal to Sir John Hawkins for his Decision on their respective Merits, by Quotations from their printed Anecdotes of the Doctor. Sir John hears them with uncommon Patience, and determines very properly on the Pretensions of the contending Parties.

'ALTERNATELY in Anecdotes go on;
But first begin you, Madam,' cried Sir John.
The thankful Dame low curtseyed to the Chair,
And thus, for victory panting, read the Fair:—

JOHN WOLCOT

MADAME PIOZZI

Sam Johnson was of Michael Johnson born;
Whose shop of books did Litchfield town adorn:
Wrong-headed, stubborn as a halter'd Ram;
In short, the model of our Hero Sam:
Inclined to madness too; for when his shop
Fell down, for want of cash to buy a prop,
For fear the thieves might steal the vanish'd store
He duly went each night and lock'd the door.

BOZZY

While Johnson was in Edinburgh, my Wife
To please his palate, studied for her life:
With every rarity she fill'd her house,
And gave the Doctor, for his dinner, grouse.

. . . .

MADAME PIOZZI

In Lincolnshire, a Lady showed our Friend
A Grotto, that she wish'd him to commend.
Quoth she, 'How *cool* in summer this abode!'—
'Yes, Madam,' answer'd Johnson; 'for a *toad*.'

BOZZY

Between old Scalpa's rugged isle and Rasay's,
The wind was vastly boisterous in our faces:
'Twas glorious, Johnson's figure to set sight on;
High in the boat, he looked a noble Triton.
But, lo! to damp our pleasure Fate concurs,
For Joe (the blockhead!) lost his Master's spurs:
This for the Rambler's temper was a rubber,
Who wonder'd Joseph could be such a lubber.

MADAME PIOZZI

I ask'd him if he knock'd Tom Osborne down;
As such a tale was current through the town.
Says I, 'Do tell me, Doctor, what befell.'—
'Why, dearest Lady, there is nought to tell:
I ponder'd on the properest mode to treat him;
The dog was impudent, and so I beat him.
Tom, like a fool, proclaim'd his fancied wrongs;
Others that I belaboured, held their tongues.'
Did any one, 'that he was happy,' cry;
Johnson would tell him plumply, 'twas a lie.

A Lady told him she was really so;
On which he sternly answer'd, 'Madam, no.
Sickly you are, and ugly; foolish, poor;
And therefore can't be *happy*, I am sure.
'Twould make a fellow hang himself, whose ear
Were, from *such creatures*, forced such stuff to hear.'

. . . .

BOZZY

As at Argyle's grand house my hat I took,
To seek my alehouse, thus began the Duke:
'Pray, Mister Boswell, won't you have some tea?'
To this I made my bow, and did agree.
Then to the drawing-room we both retreated,
Where Lady Betty Hamilton was seated
Close by the Duchess; who, in deep discourse,
Took no more notice of *me* than a Horse.—
Next day, *myself* and Doctor Johnson took
Our hats, to go and wait upon the Duke.
Next to himself the Duke did Johnson place;
But I, thank God, sat *second* to his Grace.
The place was due most surely to my merits;
And, faith, I was in very pretty spirits.
I plainly saw (my penetration such is),
I was not yet in favour with the Duchess.
Thought I, 'I am not disconcerted yet;
Before we part, I'll give her Grace a *sweat*.'
Then looks of intrepidity I put on,
And ask'd her if she'd have a plate of mutton.
This was a glorious deed, must be confess'd;
I knew I was the *Duke's* and not *her* guest.
Knowing (as I'm a man of tip-top breeding)
That *great folks* drink no healths while they are feeding;
I took my glass, and, looking at her Grace,
I stared her like a Devil in the face;
And in respectful terms, as was my duty,
Said I, 'My Lady Duchess, I salute ye.'
Most audible indeed was my salute,
For which some folks will say I was a Brute:
But faith, it dash'd her, as I knew it would;
But then, I knew that I was flesh and blood.

. . . .

JOHN WOLCOT

MADAME PIOZZI

Dear Doctor Johnson left off Drinks fermented;
With quarts of chocolate and cream contented:
Yet often down his throat's prodigious gutter.
Poor man! he poured a flood of melted butter.

BOZZY

With glee the Doctor did my Girl behold;
Her name Veronica, just four months old.
This name Veronica, a name though quaint,
Belonged originally to a Saint:
But to my old Great-grandam it was given,
As fine a woman as e'er went to Heaven;
And, what must add to her importance much,
This Lady's genealogy was Dutch.
The Man who did espouse this Dame divine,
Was Alexander, Earl of Kincardine;
Who poúred along my Body, like a Sluice,
The noble, noble, noble blood of Bruce:
And who that own'd this blood could well refuse
To make the World acquainted with the *news*?
But to return unto my charming Child:
About our Doctor Johnson she was wild;
And when he left off speaking, she would flutter,
Squawl for him to begin again, and sputter;
And to be near him a strong wish express'd:
Which proves he was not *such* a horrid Beast.
Her fondness for the Doctor pleased me greatly;
On which I loud exclaimed in language stately,
Nay, if I recollect aright, I *swore*,
I'd to her fortune add five hundred more.

MADAME PIOZZI

One day, as we were all in talking lost,
My Mother's favourite Spaniel stole the toast;
On which immediately I screamed, 'Fie on her.'
'Fie, Belle,' said I, 'you used to be on honour.'—
'Yes,' Johnson cried; 'but, Madam, pray be told,
The reason for the vice is, Belle grows *old*.'
But Johnson never could the Dog abide,
Because my Mother wash'd and comb'd his hide.
The truth on 't is, Belle was not too well bred,
But always would *insist* on being fed;
And very often too, the saucy Slut
Insisted upon having the *first cut*.

JOHN WOLCOT

BOZZY

Last night much care for Johnson's Cold was used,
Who hitherto without his nightcap *snooz'd*.
That nought might treat so wonderful a man ill,
Sweet Miss MacLeod did make a Cap of Flannel;
And, after putting it about his head,
She gave him Brandy as he went to bed.

. . . .

MADAME PIOZZI

The Doctor had a Cat, and christen'd Hodge,
That at his house in Fleet-street used to lodge.
This Hodge grew old, and sick; and used to wish
That all his dinners might be form'd of Fish.
To please poor Hodge, the Doctor, all so kind,
Went out, and bought him Oysters to his mind.
This every day he did; nor ask'd Black Frank,
Who deemed himself of much too high a rank,
With vulgar fish-fags to be forced to chat,
And purchase Oysters for a mangy Cat.

SIR JOHN

For God's sake stay each Anecdotic scrap;
Let me draw breath, and take a trifling nap.

RICHARD BRINSLEY SHERIDAN

1751–1816

75 *Lines by a Lady on the Loss of her Trunk*

HAVE you heard, my dear Anne, how my spirits are sunk?
Have you heard of the cause? Oh, the loss of my *Trunk*!
From exertion or firmness I've never yet slunk;
But my fortitude's gone with the loss of my *Trunk*!
Stout Lucy, my maid, is a damsel of spunk;
Yet she weeps night and day for the loss of my *Trunk*!
I'd better turn nun, and coquet with a monk;
For with whom can I flirt without aid from my *Trunk*!

Accurs'd be the thief, the old rascally hunks,
Who rifles the fair, and lays hands on their *Trunks*!
He, who robs the King's stores of the least bit of junk,
Is hang'd—while he's safe, who has plunder'd my *Trunk*!

There's a phrase amongst lawyers, when *nunc*'s put for *tunc*;
But, tunc and nunc both, must I grieve for my *Trunk*!
Huge leaves of that great commentator, old Brunck,
Perhaps was the paper that lin'd my poor *Trunk*!
But my rhymes are all out;—for I dare not use st——k;
'Twou'd shock Sheridan more than the loss of my *Trunk*.

76 *On Lady Anne Hamilton*

PRAY how did she look? Was she pale, was she wan?
She was blooming and red as a cherry—poor Anne.

Did she eat? Did she drink? Yes, she drank up a can,
And ate very near a whole partridge—poor Anne.

Pray what did she do? Why, she talked to each man
And flirted with Morpeth and Breanbie—poor Anne.

Pray how was she drest? With a turban and fan,
With ear-rings, with chains, and with bracelets—poor Anne.

And how went she home? In a good warm sedan
With a muff and a cloak and a tippet—poor Anne.

WILLIAM BLAKE

1757–1827

77 *Imitation of Pope: A Compliment to the Ladies*

WONDROUS the Gods, more wondrous are the Men,
More wondrous wondrous still the Cock & Hen,
More wondrous still the Table, Stool & Chair;
But Ah! More wondrous still the Charming Fair.

78

An Epitaph

I WAS buried near this Dyke,
That my friends may weep as much as they like.

ROBERT BURNS

1759–1796

79

from *The Jolly Beggars: A Cantata*

RECITATIVO

Poor Merry-Andrew, in the neuk,
  Sat guzzling wi' a tinkler-hizzie;
They mind't na wha the chorus teuk,
  Between themsels they were sae busy:
  At length, wi' drink an' courting dizzy,
He stoiter'd up an' made a face;
  Then turn'd an' laid a smack on Grizzie,
Syne tun'd his pipes wi' grave grimace.

AIR

Tune: 'Auld Sir Symon'

Sir Wisdom's a fool when he's fou;
  Sir Knave is a fool in a session;
He's there but a prentice I trow,
  But I am a fool by profession.

My grannie she bought me a beuk,
  An' I held awa to the school;
I fear I my talent misteuk,
  But what will ye hae of a fool?

For drink I would venture my neck;
  A hizzie's the half of my craft;
But what could ye other expect
  Of ane that's avowedly daft?

neuk] corner *tinkler-hizzie*] tinker hussy *mind't na*] cared not
stoiter'd] staggered *smack*] kiss *Syne*] then *fou*] drunk *beuk*] book

I ance was tied up like a stirk,
For civilly swearing and quaffin;
I ance was abus'd i' the kirk,
For towsing a lass i' my daffin.

Poor Andrew that tumbles for sport,
Let naebody name wi' a jeer;
There's ev'n, I'm tauld, i' the Court
A tumbler ca'd the Premier.

Observ'd ye yon reverend lad
Mak faces to tickle the mob;
He rails at our mountebank squad,—
It's rivalship just i' the job.

And now my conclusion I'll tell,
For faith I'm confoundedly dry;
The chiel that's a fool for himsel',
Guid Lord! he's far dafter than I.

80 *Holy Willie's Prayer*

And send the Godly in a pet to pray—POPE

Argument

Holy Willie was a rather oldish bachelor Elder in the parish of Mauchline, & much & justly famed for that polemical chattering which ends in tippling Orthodoxy, & for that Spiritualised Bawdry which refines to Liquorish Devotion.—In a Sessional process with a gentleman in Mauchline, a Mr Gavin Hamilton, Holy Willie, & his priest, father Auld, after full hearing in the Presbytery of Ayr, came off but second best; owing partly to the oratorical powers of Mr Robt Aiken, Mr Hamilton's Counsel; but chiefly to Mr Hamilton's being one of the most irreproachable & truly respectable characters in the country. On losing his Process, the Muse overheard him at his devotions as follows—

O THOU that in the heavens does dwell!
Wha, as it pleases best thysel,
Sends ane to heaven & ten to hell,
A' for thy glory!
And no for ony gude or ill
They've done before thee.

ance] once *stirk*] bullock *towsing*] tumbling *daffin*] petting *chiel*] chap

I bless & praise thy matchless might,
When thousands thou has left in night,
That I am here before thy sight,
For gifts & grace,
A burning & a shining light
To a' this place.

What was I, or my generation,
That I should get such exaltation?
I, wha deserv'd most just damnation,
For broken laws
Sax thousand years ere my creation,
Thro' Adam's cause.

When from my mother's womb I fell,
Thou might hae plunged me deep in hell,
To gnash my gooms, & weep, & wail,
In burning lakes,
Where damned devils roar & yell
Chain'd to their stakes.

Yet I am here, a chosen sample,
To shew thy grace is great & ample:
I'm here, a pillar o' thy temple
Strong as a rock,
A guide, a ruler & example
To a' thy flock.

But yet—O Lord—confess I must—
At times I'm fash'd wi' fleshly lust;
And sometimes too, in warldly trust
Vile Self gets in;
But thou remembers we are dust,
Defil'd wi' sin.

O Lord—yestreen—thou kens—wi' Meg—
Thy pardon I sincerely beg!
O may't ne'er be a living plague,
To my dishonor!
And I'll ne'er lift a lawless leg
Again upon her.

gooms] gums *fash'd*] troubled *yestreen*] last night *kens*] knowest

Besides, I farther maun avow,
Wi' Leezie's lass, three times—I trow—
But, Lord, that friday I was fou
When I cam near her;
Or else, thou kens, thy servant true
Wad never steer her.

Maybe thou lets this fleshly thorn
Buffet thy servant e'en & morn,
Lest he o'er proud & high should turn,
That he's sae gifted;
If sae, thy hand maun e'en be borne
Untill thou lift it.

Lord bless thy Chosen in this place,
For here thou has a chosen race:
But God, confound their stubborn face,
And blast their name,
Wha bring thy rulers to disgrace
And open shame.

Lord mind Gaun Hamilton's deserts!
He drinks, & swears, & plays at cartes,
Yet has sae mony taking arts
Wi' great & sma',
Frae God's ain priest the people's hearts
He steals awa.

And when we chasten'd him therefore,
Thou kens how he bred sic a splore,
And set the warld in a roar
O' laughin at us:
Curse thou his basket and his store,
Kail & potatoes.

Lord hear my earnest cry & prayer
Against that Presbytry of Ayr!
Thy strong right hand, Lord, make it bare
Upon their heads!
Lord visit them, & dinna spare,
For their misdeeds!

maun] must *Wad*] would *steer*] meddle with *sae*] so *rulers*] elders
mind] remember *cartes*] cards *ain*] own *splore*] fuss *Kail*] cabbage

O Lord my God, that glib-tongu'd Aiken!
My very heart & flesh are quaking
To think how I sat, sweating, shaking,
And piss'd wi' dread,
While Auld wi' hingin lip gaed sneaking
And hid his head!

Lord, in thy day o' vengeance try him!
Lord, visit him that did employ him!
And pass not in thy mercy by them;
Nor hear their prayer;
But for thy people's sake destroy them,
And dinna spare!

But Lord; remember me & mine
Wi' mercies temporal & divine!
That I for grace & gear may shine,
Excell'd by nane!
And a' the glory shall be thine!
Amen! Amen!

81 *The Deil's Awa wi' th' Exciseman*

THE deil cam fiddlin thro' the town,
And danc'd awa wi' th' Exciseman;
And ilka wife cries, auld Mahoun,
I wish you luck o' the prize, man.

The deil's awa, the deil's awa
The deil's awa wi' th' Exciseman,
He's danc'd awa, he's danc'd awa
He's danc'd awa wi' th' Exciseman.

We'll mak our maut and we'll brew our drink,
We'll laugh, sing, and rejoice, man;
And mony braw thanks to the meikle black deil,
That danc'd awa wi' th' Exciseman.

80 *hingin*] hanging *gaed*] went *gear*] wealth

81 *ilka*] every *Mahoun*] devil *maut*] malt *braw*] hearty *meikle*] big

There's threesome reels, there's foursome reels,
There's hornpipes and strathspeys, man,
But the ae best dance e'er cam to the Land
Was, the deil's awa wi' th' Exciseman.

CATHERINE FANSHAWE

1765–1834

82 *Enigma*

'TWAS whispered in Heaven, 'twas muttered in Hell,
And echo caught softly the sound as it fell:
In the confines of earth 'twas permitted to rest,
And the depth of the ocean its presence confessed;
'Twas seen in the lightning, 'twas heard in the thunder,
'Twill be found in the spheres when they're riven asunder;
'Twas given to man with his earliest breath,
It assists at his birth and attends him in death,
Presides o'er his happiness, honour, and health,
'Tis the prop of his house and the end of his wealth;
It begins every hope, every wish it must bound,
With the husbandman toils, and with monarchs is crowned;
In the heaps of the miser 'tis hoarded with care,
But is sure to be lost in the prodigal heir;
Without it the soldier and sailor may roam,
But woe to the wretch who expels it from home;
In the whispers of conscience it there will be found,
Nor e'er in the whirlwind of passion be drowned;
It softens the heart, and though deaf to the ear,
It will make it acutely and instantly hear;
But in shades let it rest, like an elegant flower,
Oh! breathe on it softly, it dies in an hour.

RICHARD ALFRED MILLIKIN

1767–1815

83

The Groves of Blarney

THE groves of Blarney they look so charming,
Down by the purling of sweet, silent streams,
Being banked with posies that spontaneous grow there,
Planted in order by the sweet rock close.
'Tis there's the daisy and the sweet carnation,
The blooming pink and the rose so fair,
The daffodowndilly, likewise the lily,
All flowers that scent the sweet, fragrant air.

'Tis Lady Jeffers that owns this station;
Like Alexander, or Queen Helen fair,
There's no commander in all the nation,
For emulation, can with her compare.
Such walls surround her, that no nine-pounder
Could dare to plunder her place of strength;
But Oliver Cromwell he did her pommell,
And made a breach in her battlement.

There's gravel walks there for speculation
And conversation in sweet solitude.
'Tis there the lover may hear the dove, or
The gentle plover in the afternoon;
And if a lady would be so engaging
As to walk alone in those shady bowers,
'Tis there the courtier he may transport her
Into some fort, or all under ground.

'Tis there's the kitchen hangs many a flitch in
With the maids a stiching upon the stair;
The bread and biske', the beer and whisky,
Would make you frisky if you were there.
'Tis there you'd see Peg Murphy's daughter
A washing praties forenent the door,
With Roger Cleary, and Father Healy,
All blood relations to my Lord Donoughmore.

For 'tis there's a cave where no daylight enters,
But cats and badgers are for ever bred;
Being mossed by nature, that makes it sweeter
Than a coach-and-six or a feather bed.
'Tis there the lake is, well stored with perches,
And comely eels in the verdant mud;
Besides the leeches, and groves of beeches,
Standing in order for to guard the flood.

There's statues gracing this noble place in—
All heathen gods and nymphs so fair;
Bold Neptune, Plutarch, and Nicodemus,
All standing naked in the open air!
So now to finish this brave narration,
Which my poor geni' could not entwine;
But were I Homer, or Nebuchadnezzar,
'Tis in every feature I would make it shine.

ANONYMOUS

84

The Rakes of Mallow

BEAUING, belling, dancing, drinking,
Breaking windows, damning, sinking,
Ever raking, never thinking,
        Live the rakes of Mallow.

Spending faster than it comes,
Beating waiters, bailiffs, duns,
Bacchus' true-begotten sons,
        Live the rakes of Mallow.

One time naught but claret drinking,
Then like politicians thinking
To raise the sinking-funds when sinking,
        Live the rakes of Mallow.

When at home with dadda dying
Still for Mallow water crying;
But where there's good claret plying,
        Live the rakes of Mallow.

Living short but merry lives;
Going where the devil drives;
Having sweethearts but no wives,
    Live the rakes of Mallow.

Racking tenants, stewards teasing,
Swiftly spending, slowly raising,
Wishing to spend all their lives in
    Raking as in Mallow.

Then to end this raking life,
They get sober, take a wife,
Ever after live in strife,
    And wish again for Mallow.

JOHN HOOKHAM FRERE

1769–1846

85 from *Whistlecraft*

[*King Arthur's Court*]

AND certainly they say, for fine behaving
  King Arthur's Court has never had its match;
True point of honour, without pride or braving,
  Strict etiquette for ever on the watch:
Their manners were refin'd and perfect—saving
  Some modern graces, which they could not catch,
As spitting through the teeth, and driving stages,
Accomplishments reserv'd for distant ages.

They look'd a manly, generous generation;
  Beards, shoulders, eyebrows, broad, and square, and thick,
Their accents firm and loud in conversation,
  Their eyes and gestures eager, sharp, and quick,
Shew'd them prepar'd, on proper provocation,
  To give the lie, pull noses, stab and kick;
And for that very reason, it is said,
They were so very courteous and well-bred.

The ladies look'd of an heroic race—
  At first a general likeness struck your eye,
Tall figures, open features, oval face,
  Large eyes, with ample eyebrows arch'd and high;
Their manners had an odd, peculiar grace,
  Neither repulsive, affable, nor shy,
Majestical, reserv'd, and somewhat sullen;
Their dresses partly silk, and partly woollen.

GEORGE CANNING

1770–1827

86 *The Friend of Humanity and the Knife-Grinder*

FRIEND OF HUMANITY

'NEEDY Knife-grinder! whither are you going?
Rough is the road, your Wheel is out of order—
Bleak blows the blast;—your hat has got a hole in't,
          So have your breeches!

'Weary Knife-grinder! little think the proud ones
Who in their coaches roll along the turnpike-
Road, what hard work 'tis crying all day, "Knives and
          Scissors to grind O!"

'Tell me, Knife-grinder, how came you to grind knives?
Did some rich man tyrannically use you?
Was it the Squire? or Parson of the parish?
          Or the Attorney?

'Was it the Squire for killing of his game? or
Covetous Parson, for his Tithes distraining?
Or roguish Lawyer, made you lose your little
          All in a law-suit?

'(Have you not read the Rights of Man, by Tom Paine?)
Drops of compassion tremble on my eyelids
Ready to fall, as soon as you have told your
          Pitiful story.'

KNIFE-GRINDER

'Story! God bless you! I have none to tell, Sir,
Only last night a-drinking at the Chequers,
This poor old hat and breeches, as you see, were
Torn in a scuffle.

'Constables came up for to take me into
Custody; they took me before the justice;
Justice Oldmixon put me in the Parish-
stocks for a Vagrant.

'I should be glad to drink your Honour's health in
A pot of beer, if you will give me sixpence;
But for my part, I never love to meddle
With Politics, sir.'

FRIEND OF HUMANITY

'*I* give thee sixpence! I will see thee damn'd first—
Wretch! whom no sense of wrongs can rouse to vengeance—
Sordid, unfeeling, reprobate, degraded,
Spiritless outcast!'

[*Kicks the Knife-grinder, overturns his wheel, and exit in a transport of republican enthusiasm and universal philanthropy*]

87 from *The Rovers*, a Romantic Drama

[*Rogero's Song*]

WHENE'ER with haggard eyes I view
This Dungeon, that I'm rotting in,
I think of those Companions true
Who studied with me at the U—
—niversity of Gottingen—
—niversity of Gottingen.

[*Weeps, and pulls out a blue kerchief, with which he wipes his eyes; gazing tenderly at it, he proceeds*—]

Sweet kerchief, check'd with heav'nly blue,
Which once my love sat knotting in!—
Alas! MATILDA *then* was true!—
At least I thought so at the U—
—niversity of Gottingen—
—niversity of Gottingen.

[*At the repetition of this line* ROGERO *clanks his Chains in cadence*]

Barbs! Barbs! alas! how swift you flew
 Her neat Post-Waggon trotting in!
Ye bore MATILDA from my view.
Forlorn I languish'd at the U—
 —niversity of Gottingen—
 —niversity of Gottingen.

This faded form! this pallid hue!
 This blood my veins is clotting in.
My years are many—They were few
When first I entered at the U—
 —niversity of Gottingen—
 —niversity of Gottingen.

There first for thee my passion grew,
 Sweet! sweet MATILDA POTTINGEN!
Thou wast the daughter of my Tu—
—tor, *Law Professor* at the U—
 —niversity of Gottingen!
 —niversity of Gottingen!

(Sun, moon, and thou vain world, adieu!
 That kings and priests are plotting in:
Here, doomed to starve on water gru—
—el, never shall I see the U—
 —niversity of Gottingen,
 —niversity of Gottingen.)

[*During the last stanza,* ROGERO *dashes his head repeatedly against the walls of his Prison; and, finally, so hard as to produce a visible contusion. He then throws himself on the floor in agony. The Curtain drops—the Music still continuing to play, till it is wholly fallen*]

Barbs] barbary horses

SYDNEY SMITH

1771–1845

88

Recipe for a Salad

To make this condiment, your poet begs
The pounded yellow of two hard-boiled eggs;
Two boiled potatoes, passed through kitchen-sieve,
Smoothness and softness to the salad give;
Let onion atoms lurk within the bowl,
And, half-suspected, animate the whole.
Of mordant mustard add a single spoon,
Distrust the condiment that bites so soon;
But deem it not, thou man of herbs, a fault,
To add a double quantity of salt.
And, lastly, o'er the flavored compound toss
A magic soup-spoon of anchovy sauce.
Oh, green and glorious! Oh, herbaceous treat!
'T would tempt the dying anchorite to eat;
Back to the world he'd turn his fleeting soul,
And plunge his fingers in the salad bowl!
Serenely full, the epicure would say,
Fate can not harm me, I have dined to-day!

CHARLES LAMB

1775–1834

89

Nonsense Verses

Lazy-bones, lazy-bones, wake up and peep!
The cat's in the cupboard, your mother's asleep.
There you sit snoring, forgetting her ills;
Who is to give her her Bolus and Pills?
Twenty fine Angels must come into town,
All for to help you to make your new gown:
Dainty aerial Spinsters and Singers;
Aren't you ashamed to employ such white fingers?

Delicate hands, unaccustom'd to reels,
To set 'em working a poor body's wheels?
Why they came down is to me all a riddle,
And left Hallelujah broke off in the middle;
Jove's court, and the Presence angelical, cut—
To eke out the work of a lazy young slut.
Angel-duck, Angel-duck, winged and silly,
Pouring a watering-pot over a lily,
Gardener gratuitous, careless of pelf,
Leave her to water her lily herself,
Or to neglect it to death if she chuse it:
Remember the loss is her own, if she lose it.

WALTER SAVAGE LANDOR

1775–1864

90

The Georges

GEORGE the First was always reckoned
Vile, but viler George the Second;
And what mortal ever heard
Any good of George the Third?
When from earth the Fourth descended
(God be praised!) the Georges ended.

91

Ireland

IRELAND never was contented.
Say you so? You are demented.
Ireland was contented when
All could use the sword and pen,
And when Tara rose so high
That her turrets split the sky,
And about her courts were seen
Liveried angels robed in green,
Wearing, by St Patrick's bounty,
Emeralds big as half the county.

THOMAS MOORE

1779–1852

92 *The Duke Is the Lad*

THE Duke is the lad to frighten a lass,
Galloping, dreary duke;
The Duke is the lad to frighten a lass,
He's an ogre to meet, and the d——l to pass,
With his charger prancing,
Grim eye glancing,
Chin, like a Mufti,
Grizzled and tufty,
Galloping, dreary Duke.

Ye misses, beware of the neighbourhood
Of this galloping dreary Duke;
Avoid him, all who see no good
In being run o'er by a Prince of the Blood.
For, surely, no nymph is
Fond of a grim phiz,
And of the married,
Whole crowds have miscarried
At sight of this dreary Duke.

93 from *The Fudge Family in Paris*

[*Letter III: From Mr Bob Fudge to Richard ——, Esq.*]

OH Dick! You may talk of your writing and reading,
Your Logic and Greek, but there's nothing like feeding;
And *this* is the place for it, DICKY, you dog,
Of all places on earth—the head-quarters of Prog!
Talk of England—her famed Magna Charta, I swear, is
A humbug, a flam, to the Carte at old VÉRY'S;
And as for your Juries—*who* would not set o'er 'em
A Jury of Tasters, with woodcocks before 'em?
Give CARTWRIGHT his Parliaments, fresh every year;
But those friends of *short Commons* would never do here;
And, let ROMILLY speak as he will on the question,
No Digest of Law's like the laws of digestion!

By the by, DICK, *I* fatten—but *n'importe* for that,
'Tis the mode—your Legitimates always get fat.
There's the R——G——T, there's LOUIS—and BONEY tried too,
But, tho' somewhat imperial in paunch, 'twouldn't do:—
He improv'd, indeed, much in this point, when he wed,
But he ne'er grew right royally fat *in the head.*

DICK, DICK, what a place is this Paris!—but stay—
As my raptures may bore you, I'll just sketch a Day,
As we pass it, myself and some comrades I've got,
All thorough-bred *Gnostics*, who know what is what.

After dreaming some hours of the land of Cocaigne,
    That Elysium of all that is *friand* and nice,
Where for hail they have *bon-bons*, and claret for rain,
    And the skaters in winter show off on *cream*-ice;
Where so ready all nature its cookery yields,
Macaroni au parmesan grows in the fields;
Little birds fly about with the true pheasant taint,
And the geese are all born with a liver complaint!
I rise—put on neck-cloth—stiff, tight, as can be—
For a lad who *goes into the world*, DICK, like me,
Should have his neck tied up, you know—there's no doubt of it—
Almost as tight as *some* lads who *go out of it.*
With whiskers well oil'd, and with boots that 'hold up
'The mirror to nature'—so bright you could sup
Off the leather like china; with coat, too, that draws
On the tailor, who suffers, a martyr's applause!—
With head bridled up, like a four-in-hand leader,
And stays—devil's in them—too tight for a feeder,
I strut to the old Café Hardy, which yet
Beats the field at a *déjeûner à la fourchette.*
There, DICK, what a breakfast!—oh, not like your ghost
Of a breakfast in England, your curst tea and toast;
But a side-board, you dog, where one's eye roves about,
Like a Turk's in the Haram, and thence singles out
One's *paté* of larks, just to tune up the throat,
One's small limbs of chickens, done *en papillote*,
One's erudite cutlets, drest all ways but plain,
Or one's kidneys—imagine, DICK—done with champagne!
Then, some glasses of *Beaune*, to dilute—or, mayhap,
Chambertin, which you know's the pet tipple of NAP,
And which Dad, by the by, that legitimate stickler,
Much scruples to taste, but *I*'m not so partic'lar.—

Your coffee comes next, by prescription: and then, DICK, 's
The coffee's ne'er-failing and glorious appendix,
A neat glass of *parfait-amour*, which one sips
Just as if bottled velvet tipp'd over one's lips.
This repast being ended, and *paid for*—(how odd!
  Till a man's us'd to paying, there's something so queer in't!)—
The sun now well out, and the girls all abroad,
  And the world enough air'd for us, Nobs, to appear in't,
We lounge up the Boulevards, where—oh, DICK, the phyzzes,
The turn-outs, we meet—what a nation of quizzes!
Here toddles along some old figure of fun,
With a coat you might date Anno Domini I;
A lac'd hat, worsted stockings, and—noble old soul!
A fine ribbon and cross in his best button-hole;
Just such as our PR——CE, who nor reason nor fun dreads,
Inflicts, without ev'n a court-martial, on hundreds.
Here trips a *grisette*, with a fond, roguish eye,
(Rather eatable things these *grisettes* by the by);
And there an old *demoiselle*, almost as fond,
In a silk that has stood since the time of the Fronde.
There goes a French Dandy—ah, DICK! unlike some ones
We've seen about WHITE'S—the Mounseers are but rum ones;
Such hats!—fit for monkies—I'd back Mrs DRAPER
To cut neater weather-boards out of brown paper:
And coats—how I wish, if it wouldn't distress 'em,
They'd club for old BR——MM——L, from Calais, to dress 'em!
The collar sticks out from the neck such a space,
  That you'd swear 'twas the plan of this head-lopping nation,
To leave there behind them a snug little place
  For the head to drop into, on decapitation.
In short, what with mountebanks, counts, and friseurs,
Some mummers by trade, and the rest amateurs—
What with captains in new jockey-boots and silk breeches,
  Old dustmen with swinging great opera-hats,
And shoeblacks reclining by statues in niches,
  There never was seen such a race of Jack Sprats!

From the Boulevards—but hearken!—yes—as I'm a sinner,
The clock is just striking the half-hour to dinner:
So *no* more at present—short time for adorning—
My Day must be finish'd some other fine morning.
Now, hey for old BEAUVILLIERS' larder, my boy!
And, once *there*, if the Goddess of Beauty and Joy
Were to write 'Come and kiss me, dear BOB!' I'd not budge—
Not a step, DICK, as sure as my name is

R. FUDGE

94 *Announcement of a New Grand Acceleration Company for the Promotion of the Speed of Literature*

LOUD complaints being made, in these quick-reading times,
Of too slack a supply, both of prose works and rhymes,
A new Company, form'd on the keep-moving plan,
First propos'd by the great firm of Catch-'em-who-can,
Beg to say they've now ready, in full wind and speed,
Some fast-going authors, of quite a new breed—
Such as not he who *runs* but who *gallops* may read—
And who, if well curried and fed, they've no doubt,
Will beat even Bentley's swift stud out and out.

In fact, there's no saying, so gainful the trade,
How fast immortalities now may be made;
Since Helicon never will want an 'Undying One,'
As long as the public continues a Buying One;
And the company hope yet to witness the hour,
When, by strongly applying the mare-motive power,
A three-decker novel, 'midst oceans of praise,
May be written, launch'd, read, and—forgot, in three days!

In addition to all this stupendous celerity,
Which—to the no small relief of posterity—
Pays off at sight the whole debit of fame,
Nor troubles futurity ev'n with a name,
We, the Company—still more to show how immense
Is the power o'er the mind of pounds, shillings, and pence,
Beg to add, as our literature soon may compare,
In its quick make and vent, with our Birmingham ware,
And it doesn't at all matter in either of these lines,
How *sham* is the article, so it but *shines*,—
We keep authors ready, all perch'd, pen in hand,
To write off, in any giv'n style, at command.
No matter what bard, be he living or dead,
Ask a work from his pen, and 'tis done soon as said:
There being, on th' establishment, six Walter Scotts,
One capital Wordsworth, and Southeys in lots;—
Three choice Mrs Nortons, all singing like syrens,
While most of our pallid young clerks are Lord Byrons.

. . . .

Bentley's] Richard Bentley the publisher *mare-motive power*] ''tis money makes the mare to go'

The company, since they set up in this line,
Have mov'd their concern, and are now at the sign
Of the Muse's Velocipede, *Fleet* Street, where all
Who wish well to the scheme are invited to call.

WILLIAM HONE

1780–1842

95 from *The Political House that Jack Built*

[*The Prince Regent*]

THIS is THE MAN—all shaven and shorn,
All cover'd with Orders—and all forlorn;
THE DANDY OF SIXTY,
who bows with a grace,
And has *taste* in wigs, collars,
cuirasses and lace;
Who, to tricksters, and fools,
leaves the State and its treasure,
And, when Britain's in tears,
sails about at his pleasure:
Who spurn'd from his presence
the Friends of his youth,
And now has not one
who will tell him the truth;
Who took to his counsels,
in evil hour,
The Friends to the Reasons
of lawless Power;
That back the Public Informer,
who
Would put down the *Thing*,
that, in spite of new Acts,
And attempts to restrain it,
by Soldiers or Tax,
Will *poison* the Vermin,
That plunder the Wealth,
That lay in the House,
That Jack built.

the *Thing*] the printing-press

ANONYMOUS

96

Queen Caroline

MOST Gracious Queen, we thee implore
To go away and sin no more,
But if the effort be too great,
To go away at any rate.

JANE TAYLOR

1783–1824

97

Recreation

—WE took our work, and went, you see,
To take an early cup of tea.
We did so now and then, to pay
The friendly debt, and so did they.
Not that our friendship burnt so bright
That all the world could see the light;
'Twas of the ordinary *genus*,
And little love was lost between us:
We lov'd, I think, about as true
As such near neighbours mostly do.

At first, we all were somewhat dry;
Mamma felt cold, and so did I:
Indeed, that room, sit where you will,
Has draught enough to turn a mill.
'I hope you're warm,' says Mrs G.
'O, quite so,' says mamma, *says she*;
'I'll take my shawl off by and by.'—
'This room is always warm,' *says I*.

At last the tea came up, and so,
With that, our tongues began to go.
Now, in that house you're sure of knowing
The smallest scrap of news that's going;
We find it *there* the wisest way
To take some care of what we say.

—Says she, 'there's dreadful doings still
In that affair about the *will*;
For now the folks in Brewer's Street
Don't speak to *James's*, when they meet.
Poor Mrs *Sam* sits all alone,
And frets herself to skin and bone.
For months she manag'd, she declares,
All the old gentleman's affairs;
And always let him have his way,
And never left him night nor day;
Waited and watch'd his every look,
And gave him every drop he took.
Dear Mrs *Sam*, it was too bad!
He might have left her all he had.'

'Pray ma'am,' says I, 'has poor Miss A.
Been left as *handsome* as they say?'
'My dear,' says she, ''tis no such thing,
She'd nothing but a mourning ring.
But is it not *uncommon* mean
To wear that rusty bombazeen!'
'She had,' says I, 'the very same
Three years ago, for—what's his name?'—
'The Duke of *Brunswick*,—very true,
And has not bought a thread of new,
I'm positive,' said Mrs G.—
So then we laugh'd, and drank our tea.

'So,' says mamma, 'I find it's true
What Captain P. intends to do;
To hire that house, or else to buy—'
'Close to the tan-yard, ma'am,' says I;
'Upon my word it's very strange,
I wish they mayn't repent the change!'
'My dear,' says she, ''tis very well
You know, if *they* can bear the smell.'
'Miss F.' says I, 'is said to be
A sweet young woman, is not she?'
'O, excellent! I hear,' she cried;
'O, truly so!' mamma replied.
'How old should you suppose her, pray?
She's older than she looks, they say.'
'Really,' says I, 'she seems to me
Not more than twenty-two or three.'

'O, then you're wrong,' says Mrs G.
'Their upper servant told our *Jane*,
She'll not see twenty-nine again.'
'Indeed, so old! I wonder why
She does not marry, then,' says I;
'So many thousands to bestow,
And such a beauty, too, you know.'
'A beauty! O, my dear Miss B.
You must be joking now,' says she;
Her *figure's* rather pretty,'—'Ah!
That's what *I* say,' replied mamma.

'Miss F.' says I, 'I've understood,
Spends all her time in doing good:
The people say her coming down
Is quite a blessing to the town.'
At that our hostess fetch'd a sigh,
And shook her head; and so, says I,
'It's very kind of her, I'm sure,
To be so generous to the poor.'
'No doubt,' says she, ''tis very true;
Perhaps there may be *reasons* too:—
You know some people like to pass
For *patrons* with the lower class.'

And here I break my story's thread,
Just to remark, that what she said,
Although I took the other part,
Went like a cordial to my heart.

Some innuendos more had pass'd,
Till out the scandal came at last.
'Come then, I'll tell you something more,'
Says she,—'Eliza, shut the door.—
I would not trust a creature here,
For all the world, but you, my dear.
Perhaps it's false—I wish it may,
—But let it go no further, pray!'
'O,' says mamma, 'You need not fear,
We never mention what we hear.'
And so, we drew our chairs the nearer,
And whispering, lest the child should hear her,
She told a tale, at least too *long*
To be repeated in a song;

We, panting every breath between,
With curiosity and spleen.
And how we did enjoy the sport!
And echo every faint report,
And answer every candid doubt,
And turn her motives inside out,
And holes in all her virtues pick,
Till we were sated, almost sick.

—Thus having brought it to a close,
In great good-humour, we arose.
Indeed, 'twas more than time to go,
Our boy had been an hour below.
So, warmly pressing Mrs G.
To fix a day to come to tea,
We muffled up in cloke and plaid,
And trotted home behind the lad.

THOMAS LOVE PEACOCK

1785–1866

98

The Ghosts

In life three ghostly friars were we,
And now three friarly ghosts we be.
Around our shadowy table placed,
The spectral bowl before us floats:
With wine that none but ghosts can taste,
We wash our unsubstantial throats.
Three merry ghosts—three merry ghosts—three merry ghosts are we:
Let the ocean be Port, and we'll think it good sport
To be laid in that Red Sea.

With songs that jovial spectres chaunt,
Our old refectory still we haunt.
The traveller hears our midnight mirth:
'O list!' he cries, 'the haunted choir!
'The merriest ghost that walks the earth,
'Is sure the ghost of a ghostly friar.'

Three merry ghosts—three merry ghosts—three merry
 ghosts are we:
Let the ocean be Port, and we'll think it good sport
To be laid in that Red Sea.

GEORGE GORDON, LORD BYRON

1788–1824

99 *Lines to Mr Hodgson*

Written on Board the Lisbon Packet

HUZZA! Hodgson, we are going,
 Our embargo's off at last;
Favourable breezes blowing
 Bend the canvas o'er the mast.
From aloft the signal's streaming,
 Hark! the farewell gun is fired;
Women screeching, tars blaspheming.
 Tell us that our time's expired.
  Here's a rascal
  Come to task all,
 Prying from the custom-house;
  Trunks unpacking,
  Cases cracking,
 Not a corner for a mouse
'Scapes unsearch'd amid the racket,
Ere we sail on board the Packet.

Now our boatmen quit their mooring.
 And all hands must ply the oar;
Baggage from the quay is lowering,
 We're impatient, push from shore.
'Have a care! that case holds liquor—
 Stop the boat—I'm sick—oh Lord!'
'Sick, ma'am, damme, you'll be sicker
 Ere you've been an hour on board.'
  Thus are screaming
  Men and women,

Gemmen, ladies, servants, Jacks;
Here entangling,
All are wrangling,
Stuck together close as wax.—
Such the general noise and racket,
Ere we reach the Lisbon Packet.

Now we've reach'd her, lo! the captain,
Gallant Kidd, commands the crew;
Passengers their berths are clapt in,
Some to grumble, some to spew.
'Heyday! call you that a cabin?
Why 'tis hardly three feet square:
Not enough to stow Queen Mab in—
Who the deuce can harbour there?'
'Who, sir? plenty—
Nobles twenty
Did at once my vessel fill.'—
'Did they? Jesus,
How you squeeze us!
Would to God they did so still:
Then I'd scape the heat and racket
Of the good ship, Lisbon Packet.'

Fletcher! Murray! Bob! where are you?
Stretch'd along the deck like logs—
Bear a hand, you jolly tar, you!
Here's a rope's end for the dogs.
Hobhouse muttering fearful curses,
As the hatchway down he rolls,
Now his breakfast, now his verses,
Vomits forth—and damns our souls.
'Here 's a stanza
On Braganza—
Help!'—'A couplet?'—'No, a cup
Of warm water—'
'What's the matter?'
'Zounds! my liver's coming up;
I shall not survive the racket
Of this brutal Lisbon Packet.'

Now at length we're off for Turkey,
Lord knows when we shall come back!
Breezes foul and tempests murky
May unship us in a crack.

But, since life at most a jest is,
 As philosophers allow,
Still to laugh by far the best is,
 Then laugh on—as I do now.
  Laugh at all things,
  Great and small things,
 Sick or well, at sea or shore;
  While we're quaffing,
  Let's have laughing—
 Who the devil cares for more?—
Some good wine! and who would lack it.
Ev'n on board the Lisbon Packet?

100 *Epistle from Mr Murray to Dr Polidori*

DEAR Doctor, I have read your play,
Which is a good one in its way,—
Purges the eyes and moves the bowels,
And drenches handkerchiefs like towels
With tears, that, in a flux of grief,
Afford hysterical relief
To shatter'd nerves and quicken'd pulses,
Which your catastrophe convulses.

 I like your moral and machinery;
Your plot, too, has such scope for scenery
Your dialogue is apt and smart:
The play's concoction full of art;
Your hero raves, your heroine cries,
All stab, and everybody dies.
In short, your tragedy would be
The very thing to hear and see:
And for a piece of publication,
If I decline on this occasion,
It is not that I am not sensible
To merits in themselves ostensible,
But—and I grieve to speak it—plays
Are drugs—mere drugs, sir—now-a-days.
I had a heavy loss by 'Manuel,'—
Too lucky if it prove not annual,—

Mr Murray] John Murray, Byron's publisher *Dr Polidori*] John William Polidori, physician and author, accompanied Byron to the Continent as his secretary; he was the uncle of Dante Gabriel and Christina Rossetti

And Sotheby, with his 'Orestes,'
(Which, by the by, the author's best is,)
Has lain so very long on hand,
That I despair of all demand.
I've advertised, but see my books,
Or only watch my shopman's looks;—
Still Ivan, Ina, and such lumber,
My back-shop glut, my shelves encumber.

There's Byron too, who once did better,
Has sent me, folded in a letter,
A sort of—it's no more a drama
Than Darnley, Ivan, or Kehama:
So alter'd since last year his pen is,
I think he's lost his wits at Venice.
In short, sir, what with one and t' other,
I dare not venture on another.
I write in haste; excuse each blunder;
The coaches through the streets so thunder!
My room's so full—we've Gifford here
Reading MS, with Hookham Frere,
Pronouncing on the nouns and particles
Of some of our forthcoming Articles.

The Quarterly—Ah, sir, if you
Had but the genius to review!—
A smart critique upon St Helena,
Or if you only would but tell in a
Short compass what—but to resume:
As I was saying, sir, the room—
The room's so full of wits and bards,
Crabbes, Campbells, Crokers, Freres, and Wards,
And others, neither bards nor wits:—
My humble tenement admits
All persons in the dress of gent,
From Mr Hammond to Dog Dent.

A party dines with me to-day,
All clever men, who make their way;
Crabbe, Malcolm, Hamilton, and Chantrey,
Are all partakers of my pantry.
They're at this moment in discussion
On poor De Staël's late dissolution.

Her book, they say, was in advance—
Pray Heaven, she tell the truth of France!
Thus run our time and tongues away;—
But, to return, sir, to your play:
Sorry, sir, but I cannot deal,
Unless 'twere acted by O'Neill;
My hands so full, my head so busy,
I'm almost dead, and always dizzy;
And so, with endless truth and hurry,
Dear Doctor, I am yours,

JOHN MURRAY.

101

from *Don Juan*

(i)

[*First Love*]

YOUNG Juan wandered by the glassy brooks
Thinking unutterable things. He threw
Himself at length within the leafy nooks
Where the wild branch of the cork forest grew.
There poets find materials for their books,
And every now and then we read them through,
So that their plan and prosody are eligible,
Unless like Wordsworth they prove unintelligible.

He, Juan (and not Wordsworth), so pursued
His self-communion with his own high soul
Until his mighty heart in its great mood
Had mitigated part, though not the whole
Of its disease. He did the best he could
With things not very subject to control
And turned, without perceiving his condition,
Like Coleridge into a metaphysician.

He thought about himself and the whole earth,
Of man the wonderful and of the stars
And how the deuce they ever could have birth,
And then he thought of earthquakes and of wars,
How many miles the moon might have in girth,
Of air balloons and of the many bars
To perfect knowledge of the boundless skies.
And then he thought of Donna Julia's eyes.

In thoughts like these true wisdom may discern
  Longings sublime and aspirations high,
Which some are born with, but the most part learn
  To plague themselves withal, they know not why.
'Twas strange that one so young should thus concern
  His brain about the action of the sky.
If you think 'twas philosophy that this did,
I can't help thinking puberty assisted.

He pored upon the leaves and on the flowers
  And heard a voice in all the winds; and then
He thought of wood nymphs and immortal bowers,
  And how the goddesses came down to men.
He missed the pathway, he forgot the hours,
  And when he looked upon his watch again,
He found how much old Time had been a winner.
He also found that he had lost his dinner.

Thus would he while his lonely hours away
  Dissatisfied, nor knowing what he wanted.
Nor glowing reverie nor poet's lay
  Could yield his spirit that for which it panted,
A bosom whereon he his head might lay
  And hear the heart beat with the love it granted,
With several other things, which I forget
Or which at least I need not mention yet.

(ii)

[*Fame*]

WHAT is the end of fame? 'Tis but to fill
  A certain portion of uncertain paper.
Some liken it to climbing up a hill,
  Whose summit, like all hills, is lost in vapour.
For this men write, speak, preach, and heroes kill,
  And bards burn what they call their midnight taper,
To have, when the original is dust,
A name, a wretched picture, and worse bust.

What are the hopes of man? Old Egypt's King
  Cheops erected the first pyramid,
And largest, thinking it was just the thing
  To keep his memory whole and mummy hid;

But somebody or other rummaging,
   Burglariously broke his coffin's lid.
Let not a monument give you or me hopes,
Since not a pinch of dust remains of Cheops.

But I being fond of true philosophy
   Say very often to myself, 'Alas!
All things that have been born were born to die,
   And flesh (which Death mows down to hay) is grass.
You've passed your youth not so unpleasantly,
   And if you had it o'er again, 'twould pass;
So thank your stars that matters are no worse
And read your Bible, sir, and mind your purse.'

(iii)

[*To Be or Not to Be*]

'To be or not to be! That is the question,'
   Says Shakespeare, who just now is much in fashion.
I'm neither Alexander nor Hephaestion,
   Nor ever had for abstract fame much passion,
But would much rather have a sound digestion
   Than Buonaparte's cancer. Could I dash on
Through fifty victories to shame or fame,
Without a stomach what were a good name?

Oh *dura ilia messorum!* 'Oh
   Ye rigid guts of reapers!' I translate
For the great benefit of those who know
   What indigestion is—that inward fate
Which makes all Styx through one small liver flow.
   A peasant's sweat is worth his lord's estate.
Let this one toil for bread, that rack for rent;
He who sleeps best may be the most content.

'To be or not to be?' Ere I decide,
   I should be glad to know that which is being.
'Tis true we speculate both far and wide
   And deem because we see, we are all-seeing.
For my part, I'll enlist on neither side
   Until I see both sides for once agreeing.
For me, I sometimes think that life is death,
Rather than life a mere affair of breath.

Que sais-je? was the motto of Montaigne,
   As also of the first academicians.
That all is dubious which man may attain
   Was one of their most favourite positions.
There's no such thing as certainty; that's plain
   As any of mortality's conditions.
So little do we know what we're about in
This world, I doubt if doubt itself be doubting.

It is a pleasant voyage perhaps to float
   Like Pyrrho on a sea of speculation.
But what if carrying sail capsize the boat?
   Your wise men don't know much of navigation,
And swimming long in the abyss of thought
   Is apt to tire. A calm and shallow station
Well nigh the shore, where one stoops down and gathers
Some pretty shell, is best for moderate bathers.

'But heaven,' as Cassio says, 'is above all.
   No more of this then—let us pray!' We have
Souls to save, since Eve's slip and Adam's fall,
   Which tumbled all mankind into the grave,
Besides fish, beasts, and birds. 'The sparrow's fall
   Is special providence', though how it gave
Offence, we know not; probably it perched
Upon the tree which Eve so fondly searched.

Oh ye immortal gods, what is theogony?
   Oh thou too mortal man, what is philanthropy?
Oh world, which was and is, what is cosmogony?
   Some people have accused me of misanthropy,
And yet I know no more than the mahogany
   That forms this desk of what they mean. Lycanthropy
I comprehend, for without transformation
Men become wolves on any slight occasion.

But I, the mildest, meekest of mankind
   Like Moses or Melancthon, who have ne'er
Done anything exceedingly unkind,
   And (though I could not now and then forbear
Following the bent of body or of mind)
   Have always had a tendency to spare,
Why do they call me misanthrope? Because
They hate me, not I them. And here we'll pause.

R. H. BARHAM

1788–1845

102 ### *The Jackdaw of Rheims*

THE Jackdaw sat on the Cardinal's chair!
Bishop, and abbot, and prior were there;
  Many a monk and many a friar,
  Many a knight and many a squire,
With a great many more of lesser degree,—
In sooth a goodly company;
And they served the Lord Primate on bended knee.
  Never, I ween,
  Was a prouder seen,
Read of in books, or dreamt of in dreams,
Than the Cardinal Lord Archbishop of Rheims!

  In and out
  Through the motley rout,
That little Jackdaw kept hopping about;
  Here and there
  Like a dog in a fair,
  Over comfits and cates,
  And dishes and plates,
Cowl and cope, and rochet and pall,
Mitre and crosier! he hopped upon all!
  With saucy air,
  He perched on the chair
Where, in state, the great Lord Cardinal sat
In the great Lord Cardinal's great red hat;
  And he peered in the face
  Of his Lordship's Grace,
With a satisfied look, as if he would say,
'We two are the greatest folks here to-day!'
  And the priests with awe,
  As such freaks they saw,
Said, 'The Devil must be in that little Jackdaw!'

The feast was over, the board was cleared,
The flawns and the custards had all disappeared,
And six little Singing-boys,—dear little souls!
In nice clean faces, and nice white stoles,

Came in order due,
Two by two,
Marching that grand refectory through!
A nice little boy held a golden ewer,
Embossed and filled with water, as pure
As any that flows between Rheims and Namur,
Which a nice little boy stood ready to catch
In a fine golden hand-basin made to match.
Two nice little boys, rather more grown,
Carried lavender-water, and eau de Cologne;
And a nice little boy had a nice cake of soap,
Worthy of washing the hands of the Pope.
One little boy more
A napkin bore,
Of the best white diaper, fringed with pink,
And a Cardinal's Hat marked in 'permanent ink'.

The great Lord Cardinal turns at the sight
Of these nice little boys dressed all in white:
From his finger he draws
His costly turquoise;
And, not thinking at all about little jackdaws,
Deposits it straight
By the side of his plate,
While the nice little boys on his Eminence wait;
Till, when nobody's dreaming of any such thing,
That little Jackdaw hops off with the ring!

There's a cry and a shout,
And a deuce of a rout,
And nobody seems to know what they're about,
But the Monks have their pockets all turned inside out.
The Friars are kneeling,
And hunting and feeling
The carpet, the floor, and the walls, and the ceiling.
The Cardinal drew
Off each plum-coloured shoe,
And left his red stockings exposed to the view;
He peeps, and he feels
In the toes and the heels;
They turn up the dishes,—they turn up the plates,—
They take up the poker and poke out the grates,
They turn up the rugs,
They examine the mugs:—

But, no!—no such thing;—
They can't find THE RING!
And the Abbot declared that, 'when nobody twigged it,
Some rascal or other had popped in, and prigged it!'

The Cardinal rose with a dignified look,
He called for his candle, his bell, and his book!
In holy anger, and pious grief
He solemnly cursed that rascally thief!
He cursed him at board, he cursed him in bed;
From the sole of his foot to the crown of his head;
He cursed him in sleeping, that every night
He should dream of the devil, and wake in a fright;
He cursed him in eating, he cursed him in drinking,
He cursed him in coughing, in sneezing, in winking;
He cursed him in sitting, in standing, in lying;
He cursed him in walking, in riding, in flying,
He cursed him in living, he cursed him in dying!—
Never was heard such a terrible curse!!
But what gave rise
To no little surprise
Nobody seemed one penny the worse!
The day was gone,
The night came on
The Monks and the Friars they searched till dawn;
When the Sacristan saw,
On crumpled claw,
Coming limping a poor little lame Jackdaw!
No longer gay,
As on yesterday;
His feathers all seemed to be turned the wrong way;—
His pinions drooped—he could hardly stand,—
His head was as bald as the palm of your hand;
His eye so dim,
So wasted each limb,
That, heedless of grammar, they all cried, 'THAT'S HIM—
That's the scamp that has done this scandalous thing!
That's the thief that has got my Lord Cardinal's Ring!'
The poor little Jackdaw,
When the Monks he saw,
Feebly gave vent to the ghost of a caw;
And turned his bald head, as much as to say,
'Pray, be so good as to walk this way!'
Slower and slower

He limped on before,
Till they came to the back of the belfry door,
Where the first thing they saw,
Midst the sticks and the straw,
Was the RING in the nest of that little Jackdaw!

Then the great Lord Cardinal called for his book,
And off that terrible curse he took;
The mute expression
Served in lieu of confession,
And, being thus coupled with full restitution,
The Jackdaw got plenary absolution!
—When these words were heard,
That poor little bird
Was so changed in a moment, 'twas really absurd.
He grew sleek, and fat;
In addition to that,
A fresh crop of feathers came thick as a mat!
His tail waggled more
Even than before,
But no longer it wagged with an impudent air,
No longer he perched on the Cardinal's chair.
He hopped now about
With a gait devout;
At matins, at Vespers, he never was out;
And, so far from any more pilfering deeds,
He always seemed telling the Confessor's beads.
If any one lied,—or if any one swore,—
Or slumbered in prayer-time and happened to snore,
That good Jackdaw
Would give a great 'Caw!'
As much as to say, 'Don't do so any more!'
While many remarked, as his manners they saw,
That they 'never had known such a pious Jackdaw.'
He long lived the pride
Of that country side,
And at last in the odour of sanctity died;
When, as words were too faint
His merits to paint,
The Conclave determined to make him a Saint;
And on newly-made Saints and Popes, as you know,
It's the custom, at Rome, new names to bestow,
So they canonized him by the name of Jem Crow!

103 *Lines Left at Mr Theodore Hook's House in June, 1834*

As Dick and I
Were a-sailing by
At Fulham bridge, I cock'd my eye,
And says I, 'Add-zooks!
There's Theodore Hook's,
Whose Sayings and Doings make such pretty books.'

'I wonder,' says I,
Still keeping my eye
On the house, 'if he's in—I should like to try;'
With his oar on his knee,
Says Dick, says he,
'Father, suppose you land and see!'

'What land and *sea*,'
Says I to he.
'Together! why, Dick, why how can that be?'
And my comical son,
Who is fond of fun,
I thought would have split his sides at the pun.

So we rows to shore,
And knocks at the door—
When William, a man I've seen often before.
Makes answer and says,
'Master's gone in a chaise
Call'd a *homnibus*, drawn by a couple of bays.'

So I says then,
'Just lend me a pen:'
'I will, sir,' says William, politest of men;
So having no card, these poetical brayings,
Are the record I leave of my doings and sayings.

JOHN KEATS

1795–1821

104 *Lines Rhymed in a Letter from Oxford*

THE Gothic looks solemn,
The plain Doric column
Supports an old bishop and crosier.
The mouldering arch,
Shaded o'er by a larch,
Stands next door to Wilson the Hosier.

Vicè—that is, by turns—
O'er pale faces mourns
The black tasselled trencher and common hat.
The chantry boy sings,
The steeple-bell rings,
And as for the Chancellor—*dominat*.

There are plenty of trees,
And plenty of ease,
And plenty of fat deer for parsons.
And when it is venison,
Short is the benison,
Then each on a leg or thigh fastens.

105 from *A Song about Myself*

THERE was a naughty boy,
And a naughty boy was he,
For nothing would he do
But scribble poetry—
He took
An inkstand
In his hand,
And a pen
Big as ten
In the other.
And away
In a pother
He ran

To the mountains
And fountains
And ghostès
And postès
And witches
And ditches,
And wrote
In his coat
When the weather
Was cool—
Fear of gout—
And without
When the weather
Was warm.
Och, the charm
When we choose
To follow one's nose
To the north,
To the north,
To follow one's nose
To the north!

. . . . .

There was a naughty boy,
And a naughty boy was he,
He ran away to Scotland
The people for to see—
Then he found
That the ground
Was as hard,
That a yard
Was as long,
That a song
Was as merry,
That a cherry
Was as red,
That lead
Was as weighty,
That fourscore
Was as eighty,
That a door
Was as wooden
As in England—
So he stood in his shoes

And he wondered,
He wondered,
He stood in his shoes
And he wondered.

106

'All These Are Vile'

THE House of Mourning written by Mr Scott,
A sermon at the Magdalen, a tear
Dropped on a greasy novel, want of cheer
After a walk uphill to a friend's cot,
Tea with a maiden lady, a cursed lot
Of worthy poems with the author near,
A patron lord, a drunkenness from beer,
Haydon's great picture, a cold coffee pot
At midnight when the muse is ripe for labour,
The voice of Mr Coleridge, a French bonnet
Before you in the pit, a pipe and tabour,
A damned inseparable flute and neighbour—
All these are vile. But viler Wordsworth's sonnet
On Dover. Dover! Who *could* write upon it?

J. R. PLANCHÉ

1796–1880

107

Self-Evident

WHEN other lips and other eyes
Their tales of love shall tell,
Which means the usual sort of lies
You've heard from many a swell;
When, bored with what you feel is bosh,
You'd give the world to see
A friend whose love you know will wash,
Oh, then remember me!

Mr Scott] John Scott, editor of the *London Magazine* *the Magdalen*] a hospital for reformed prostitutes

When Signor Solo goes his tours,
And Captain Craft's at Ryde,
And Lord Fitzpop is on the moors,
And Lord knows who beside;
When to exist you feel a task
Without a friend at tea,
At such a moment I but ask
That you'll remember me.

THOMAS HAYNES BAYLY

1797–1839

108 *Out, John*

OUT, John! out, John! what are you about, John?
If you don't say 'Out,' at once, you make the fellow doubt, John!
Say I'm out, whoever calls; and hide my hat and cane, John!
Say you've not the least idea when I shall come again, John.
Let the people leave their bills, but tell them not to call, John;
Say I'm courting Miss Rupee, and mean to pay them all, John.

Run, John! run, John! there's another dun, John;
If it's Prodger, bid him call to-morrow week at one, John!
If he says he saw me at the window, as he knock'd, John!
Make a face, and shake your head, and tell him you are shocked, John!
Take your pocket-handkerchief, and put it to your eye, John!
Say your master's not the man to bid you tell a lie, John!

Oh! John, go, John! there's Noodle's knock, I know, John!
Tell him that all yesterday you sought him high and low, John!
Tell him, just before he came, you saw me mount the hill, John!
Say—you think I'm only gone to pay his little bill, John!
Then, I think, you'd better add—that if I miss to-day John!
You're sure I mean to call when next I pass his way, John!

Hie, John! fly, John! I will tell you why, John!
If there is not Grimshawe at the corner, let me die, John!
He will hear of no excuse—I'm sure he'll search the house, John!
Peeping into corners hardly fit to hold a mouse, John!
Beg he'll take a chair and wait—I know he won't refuse, John!
And I'll pop through the little door that opens on the mews, John!

THOMAS HOOD

1799–1845

109 *A First Attempt in Rhyme*

If I were used to writing verse,
And had a muse not so perverse,
But prompt at Fancy's call to spring
And carol like a bird in Spring;
Or like a Bee, in summer time,
That hums about a bed of thyme,
And gathers honey and delights
From ev'ry blossom where it 'lights;
If I, alas! had such a muse,
To touch the Reader or amuse,
And breathe the true poetic vein,
This page should not be fill'd in vain!
But ah! the pow'r was never mine
To dig for gems in Fancy's mine:
Or wander over land and main
To seek the Fairies' old domain—
To watch Apollo while he climbs
His throne in oriental climes;
Or mark the 'gradual dusky veil'
Drawn over Tempe's tuneful vale,
In classic lays remember'd long—
Such flights to bolder wings belong;
To Bards who on that glorious height,
Of sun and song, Parnassus hight,
Partake the fire divine that burns,
In Milton, Pope, and Scottish Burns,
Who sang his native braes and burns.

For me a novice strange and new,
Who ne'er such inspiration knew,
But weave a verse with travail sore,
Ordain'd to creep and not to soar,
A few poor lines alone I write,
Fulfilling thus a friendly rite,
Not meant to meet the Critic's eye,
For oh! to hope from such as I,
For anything that's fit to read,
Were trusting to a broken reed!

110 *Sonnet to Vauxhall*

THE cold transparent ham is on my fork—
 It hardly rains—and hark the bell!—ding-dingle—
Away! Three thousand feet at gravel work,
 Mocking a Vauxhall shower!—Married and Single
Crush—rush; —Soak'd Silks with wet white Satin mingle.
 Hengler! Madame! round whom all bright sparks lurk,
Calls audibly on Mr and Mrs Pringle
 To study the Sublime, &c.—(vide Burke)
All Noses are upturn'd!—Whish—ish!—On high
 The rocket rushes—trails—just steals in sight—
Then droops and melts in bubbles of blue light—
 And Darkness reigns—Then balls flare up and die—
Wheels whiz—smack crackers—serpents twist—and then
 Back to the cold transparent ham again!

111 *A Parental Ode to my Son, Aged Three Years and Five Months*

  THOU happy, happy elf!
(But stop,—first let me kiss away that tear)—
  Thou tiny image of myself!
(My love, he's poking peas into his ear!)
  Thou merry, laughing sprite!
  With spirits feather-light,
Untouch'd by sorrow and unsoil'd by sin—
(Good heavens! the child is swallowing a pin!)

  Thou little tricksy Puck!
With antic toys so funnily bestuck,
Light as the singing bird that wings the air—
(The door! the door! he'll tumble down the stair!)
  Thou darling of thy sire!
(Why, Jane, he'll set his pinafore a-fire!)
  Thou imp of mirth and joy!
In love's dear chain so strong and bright a link,
Thou idol of thy parents—(Drat the boy!
  There goes my ink!)

Thou cherub—but of earth;
Fit playfellow for Fays, by moonlight pale,
In harmless sport and mirth,
(That dog will bite him if he pulls its tail!)
Thou human humming-bee, extracting honey
From ev'ry blossom in the world that blows,
Singing in Youth's Elysium ever sunny—
(Another tumble!—that's his precious nose!)

Thy father's pride and hope!
(He'll break the mirror with that skipping-rope!)
With pure heart newly stamp'd from Nature's mint—
(Where *did* he learn that squint?)
Thou young domestic dove!
(He'll have that jug off, with another shove!)
Dear nursling of the hymeneal nest!
(Are those torn clothes his best!)
Little epitome of man!
(He'll climb upon the table, that's his plan!)
Touch'd with the beauteous tints of dawning life—
(He's got a knife!)

Thou enviable being!
No storms, no clouds, in thy blue sky foreseeing,
Play on, play on,
My elfin John!
Toss the light ball—bestride the stick—
(I knew so many cakes would make him sick!)
With fancies buoyant as the thistledown,
Prompting the face grotesque, and antic brisk,
With many a lamb-like frisk—
(He's got the scissors, snipping at your gown!)

Thou pretty opening rose!
(Go to your mother, child, and wipe your nose!)
Balmy, and breathing music like the South,
(He really brings my heart into my mouth!)
Fresh as the morn, and brilliant as its star,—
(I wish that window had an iron bar!)
Bold as the hawk, yet gentle as the dove—
(I'll tell you what, my love,
I cannot write, unless he's sent above!)

112

A Public Dinner

AT seven you just nick it,
Give card—get wine ticket;
Walk round through the Babel,
From table to table,
To find—a hard matter,—
Your name in a platter;
Your wish was to sit by
Your friend Mr Whitby,
But steward's assistance,
Has placed you at distance,
And thanks to arrangers,
You sit among strangers;
But too late for mending,—
Twelve sticks come attending
A stick of a Chairman,
A little dark spare man,
With bald shining nob,
'Mid Committee swell mob,
In short a short figure,
You thought the Duke bigger;
Then silence is wanted,
Non Nobis is chanted;
Then Chairman reads letter,
The Duke's a regretter,
A promise to break it,
But chair he can't take it;
Is grieved to be from us,
But sends friend Sir Thomas,
And what is far better,
A cheque in the letter,
Hear! hear! and a clatter,
And there ends the matter.
Now soups come and fish in,
And C—— brings a dish in;
Then rages the battle,
Knives clatter, forks rattle,
Steel forks with black handles,
Under fifty wax candles.
Your soup-plate is soon full,
You sip just a spoonful.
Mr Roe will be grateful
To send him a plateful;

And then comes the Waiter
'Must trouble for 'tater;'
And then you drink wine off
With somebody—nine off;
Bucellas, made handy,
With Cape and bad Brandy,
Or East India Sherry,
That's very hot—very.
You help Mr Myrtle,
Then find your mock turtle
Went off while you lingered
With waiter light-fingered.
To make up for gammon,
You order some salmon,
Which comes to your fauces,
With boats without sauces.
You then make a cut on
Some Lamb, big as Mutton,
And ask for some grass too,
But that you must pass too;
It serv'd the first twenty,
But toast there is plenty.
Then, while lamb gets coldish,
A goose that is oldish—
At carving not clever—
You're begg'd to dissever,
And when thus you treat it,
Find no one will eat it.
So, hungry as glutton,
You turn to your mutton,
But—no sight for laughter,
The soup it's gone after.
Mr Green then is very
Disposed to take sherry,
And then Mr Nappy
Will feel very happy,
And then Mr Conner
Requests the same honour;
Mr Clark, when at leisure,
Will really feel pleasure,
Then Waiter leans over,
To take off a cover

Bucellas] a Portuguese white wine *fauces*] the cavity at the back of the mouth

From fowls, which all beg of,
A wing or a leg of;
And while they all peck bone,
You take to a neck bone.
But even your hunger
Declares for a younger.
A fresh plate you call for,
But vainly you bawl for;
Now taste disapproves it,
No waiter removes it.
Still hope newly budding,
Relies on a pudding;
But critics each minute
Set fancy agin it—
'That's queer vermicelli.'
'I say, Vizetelly,
There's glue in that jelly.'
'Tart's bad altogether;
That crust's made of leather.'
'Some custard, friend Vesey?'
'No—batter made easy.'
'Some cheese, Mr Foster?'
'—Don't like single Glos'ter.'
Meanwhile to top table,
Like fox in the fable,
You see silver dishes,
With those little fishes,
The white bait delicious,
Borne past you officious;
And hear rather plainish,
A sound that's champaignish,
And glimpse certain bottles
Made long in the throttles,
And sniff—very pleasant!
Grouse, partridge, and pheasant,
And see mounds of ices,
For Patrons and Vices;
Pine apple, and bunches
Of grapes, for sweet munches,
And fruits of all virtue
That really *desert* you.
You've nuts, but not crack ones,
Half empty, and black ones;
With oranges sallow—

They can't be called yellow—
Some pippins well wrinkled,
And plums almond sprinkled,
Some rout cakes, and so on,
Then with business to go on;
Long speeches are stutter'd,
And toasts are well butter'd,
While dames in the gallery,
All dressed in fallallery,
Look on at the mummery:
And listen to flummery.
Hip, hip, and huzzaing,
And singing and saying,
Glees, catches, orations,
And lists of donations.
Hush, a song, Mr Tinney—
'Mr Benbow, one guinea;
Mr Frederick Manual,
One guinea, and annual.'
Song—Jockey and Jenny—
'Mr Markham, one guinea.'
'Have you all filled your glasses?
Here's a health to good lasses.'
The subscription still skinny—
'Mr Franklin, one guinea,'
Franklin looks like a ninny;
'Mr Boreham, one guinea—
Mr Brogg, Mr Finney,
Mr Tempest—one guinea,
Mr Merrington—twenty,'
Rough music in plenty.
Away toddles Chairman,
The little dark spare man
Not sorry at ending
With white sticks attending,
And some vain Tomnoddy,
Votes in his own body
To fill the void seat up,
And get on his feet up,
To say, with voice squeaking,
'Unaccustomed to speaking,'
Which sends you off seeking
Your hat, number thirty—
No coach—very dirty.

So, hungry and fever'd,
Wet-footed—spoilt-beaver'd,
Eyes aching in socket,
Ten pounds out of pocket,
To Brook-Street the Upper,
You haste home to supper.

113 *No!*

No sun—no moon!
No morn—no noon—
No dawn—no dusk—no proper time of day—
No sky—no earthly view—
No distance looking blue—
No road—no street—no 't' other side the way'—
No end to any Row—
No indications where the Crescents go—
No top to any steeple—
No recognitions of familiar people—
No courtesies for showing 'em—
No knowing 'em!—
No travelling at all—no locomotion,
No inkling of the way—no notion—
'No go'—by land or ocean—
No mail—no post—
No news from any foreign coast—
No Park—no Ring—no afternoon gentility—
No company—no nobility—
No warmth, no cheerfulness, no healthful ease,
No comfortable feel in any member—
No shade, no shine, no butterflies, no bees,
No fruits, no flowers, no leaves, no birds,—
November!

114 from *Miss Kilmansegg and her Precious Leg*

[*Her Birth*]

WHAT different dooms our birthdays bring!
For instance, one little mannikin thing
Survives to wear many a wrinkle;
While Death forbids another to wake,
And a son that it took nine moons to make,
Expires without even a twinkle!

Into this world we come like ships,
Launch'd from the docks, and stocks, and slips,
 For fortune fair or fatal;
And one little craft is cast away,
In its very first trip in Babbicome Bay,
 While another rides safe at Port Natal.

What different lots our stars accord!
This babe to be hail'd and woo'd as a Lord,
 And that to be shunned like a leper!
One, to the world's wine, honey, and corn,
Another, like Colchester native, born
 To its vinegar, only, and pepper.

One is littered under a roof
Neither wind nor water proof,—
 That's the prose of Love in a Cottage—
A puny, naked, shivering wretch,
The whole of whose birthright would not fetch,
Though Robins himself drew up the sketch,
 The bid of 'a mess of pottage.'

Born of Fortunatus's kin,
Another comes tenderly usher'd in
 To a prospect all bright and burnish'd:
No tenant he, for life's back slums—
He comes to the world as a gentleman comes
 To a lodging ready furnish'd.

And the other sex—the tender—the fair—
What wide reverses of fate are there!
While Margaret, charm'd by the Bulbul rare,
 In a garden of Gul reposes—
Poor Peggy hawks nosegays from street to street,
Till—think of that, who find life so sweet!—
 She hates the smell of roses!

Not so with the infant Kilmansegg!
She was not born to steal or beg,
 Or gather cresses in ditches;
To plait the straw or bind the shoe,
Or sit all day to hem and sew,
As females must, and not a few—
 To fill their insides with stitches!

She was not doom'd for bread to eat
To be put to her hands as well as her feet—
  To carry home linen from mangles—
Or heavy-hearted, and weary-limb'd,
To dance on a rope in a jacket trimm'd
  With as many bows as spangles.

She was one of those who by Fortune's boon
Are born, as they say, with a silver spoon
  In her mouth, not a wooden ladle:
To speak according to poet's wont,
Plutus as sponsor stood at her font,
  And Midas rock'd the cradle.

At her first *début* she found her head
On a pillow of down, in a downy bed,
  With a damask canopy over.
For although by the vulgar popular saw,
All mothers are said to be 'in the straw',
  Some children are born in clover.

Her very first thought of vital air,
It was not the common chameleon fare
  Of Plebeian lungs and noses,—
    No—her earliest sniff
    Of this world was a whiff
  Of the genuine Otto of Roses!

When she saw the light—it was no mere ray
Of that light so common—so everyday—
  That the sun each morning launches—
But six wax tapers dazzled her eyes,
From a thing—a gooseberry bush for size—
  With a golden stem and branches.

She was born exactly at half-past two,
As witness'd a timepiece in or-molu
  That stood on a marble table—
Showing at once the time of day,
And a team of *Gildings* running away
  As fast as they were able,
With a golden God with a golden Star,
And a golden spear in a golden Car
  According to Grecian fable.

Like other babes, at her birth she cried,
Which made a sensation far and wide,
Ay, for twenty miles around her;
For though to the ear 'twas nothing more
Than an infant's squall, it was really the roar
Of a Fifty-thousand Pounder!
It shook the next heir
In his library chair,
And made him cry, 'Confound her!'

Of signs and omens there was no dearth,
Any more than at Owen Glendower's birth,
Or the advent of other great people:
Two bullocks dropp'd dead,
As if knock'd on the head,
And barrels of stout
And ale ran about,
And the village-bells such a peal rang out,
That they cracked the village steeple.

In no time at all, like mushroom spawn,
Tables sprang up all over the lawn;
Not furnish'd scantly or shabbily,
But on scale as vast
As that huge repast,
With its loads and cargoes
Of drink and botargoes,
At the Birth of the Babe in Rabelais.

Hundreds of men were turn'd into beasts,
Like the guests at Circe's horrible feasts,
By the magic of ale and cider:
And each country lass, and each country lad,
Began to caper and dance like mad,
And even some old ones appear'd to have had
A bite from the Naples Spider.

Then as night came on,
It had scared King John,
Who considered such signs not risible,
To have seen the maroons,
And the whirling moons,
And the serpents of flame,
And wheels of the same,
That according to some were 'whizzable.'

Oh, happy Hope of the Kilmanseggs!
Thrice happy in head, and body, and legs,
  That her parents had such full pockets!
For had she been born of Want and Thrift,
For care and nursing all adrift,
It's ten to one she had had to make shift
  With rickets instead of rockets!

And how was the precious Baby drest?
In a robe of the East, with lace of the West,
  Like one of Croesus's issue—
    Her best bibs were made
    Of rich gold brocade,
  And the others of silver tissue.

And when the Baby inclined to nap,
She was lull'd on a Gros de Naples lap,
By a nurse, in a modish Paris cap,
  Of notions so exalted,
She drank nothing lower than Curaçoa,
Maraschino, or pink Noyau,
  And on principle never malted.

From a golden boat, with a golden spoon,
The babe was fed night, morning, and noon;
  And altho' the tale seems fabulous,
'Tis said her tops and bottoms were gilt,
Like the oats in that Stable-yard Palace built
  For the Horse of Heliogabalus.

And when she took to squall and kick,
For pain will wring, and pins will prick,
  E'en the wealthiest nabob's daughter;—
They gave her no vulgar Dalby or gin,
But a liquor with leaf of gold therein,
  Videlicet—Dantzic Water.

In short, she was born, and bred, and nurst,
And drest in the best from the very first,
  To please the genteelest censor—
And then, as soon as strength would allow,
Was vaccinated, as babes are now,
With virus ta'en from the best-bred cow
  Of Lord Althorpe's—now Earl Spencer.

ANONYMOUS

115 *Thy Heart*

THY heart is like some icy lake,
  On whose cold brink I stand;
Oh, buckle on my spirit's skate,
And lead, thou living saint, the way
  To where the ice is thin—
That it may break beneath my feet
  And let a lover in!

W. M. PRAED

1802–1839

116 *A Letter, From a Lady in London to a Lady at Lausanne*

DEAR Alice, you'll laugh when you know it,—
  Last week, at the Duchess's ball,
I danced with the clever new poet,
  You've heard of him,—Tully St Paul.
Miss Jonquil was perfectly frantic;
  I wish you had seen Lady Anne!
It really was very romantic;
  He *is* such a talented man!

He came up from Brazennose College,
  'Just caught', as they call it, last Spring;
And his head, love, is stuffed full of knowledge
  Of every conceivable thing:
Of science and logic he chatters,
  As fine and as fast as he can;
Though *I* am no judge of such matters,
  I'm sure he's a talented man.

His stories and jests are delightful;—
Not stories or jests, dear, for *you*;—
The jests are exceedingly spiteful,
The stories not always *quite* true.
Perhaps to be kind and veracious
May do pretty well at Lausanne;
But it never would answer,—good gracious!
Chez nous, in a talented man.

He sneers,—how my Alice would scold him!—
At the bliss of a sigh or a tear:
He laughed,—only think,—when I told him
How we cried o'er Trevelyan last year.
I vow I was quite in a passion;
I broke all the sticks of my fan;
But sentiment's quite out of fashion,
It seems, in a talented man.

Lady Bab, who is terribly moral,
Declared that poor Tully is vain,
And apt,—which is silly,—to quarrel,
And fond,—which is wrong—of Champagne.
I listened and doubted, dear Alice;
For I saw, when my Lady began,
It was only the Dowager's malice;
She *does* hate a talented man!

He's hideous,—I own it.—But fame, love,
Is all that these eyes can adore:
He's lame;—but Lord Byron was lame, love,
And dumpy;—but so is Tom Moore.
Then his voice,—*such* a voice! my sweet creature,
It's like your Aunt Lucy's Toucan;
But oh! what's a tone or a feature,
When once one's a talented man?

My mother, you know, all the season,
Has talked of Sir Geoffrey's estate;
And truly, to do the fool reason,
He *has* been less horrid of late.

Trevelyan] a fashionable novel of the period

But today, when we drive in the carriage,
   I'll tell her to lay down her plan;—
If ever I venture on marriage,
   It *must* be a talented man!

PS—I have found, on reflection,
   One fault in my friend,—*entre nous*;—
Without it he'd just be perfection;—
   Poor fellow,—he has not a *sou*.
And so, when he comes in September
   To shoot with my Uncle, Sir Dan,
I've promised Mamma to remember
   He's *only* a talented man!

117

Good-Night to the Season

Thus runs the world away.
HAMLET

GOOD-NIGHT to the Season! 'tis over!
   Gay dwellings no longer are gay;
The courtier, the gambler, the lover,
   Are scatter'd like swallows away:
There's nobody left to invite one,
   Except my good uncle and spouse;
My mistress is bathing at Brighton,
   My patron is sailing at Cowes:
For want of a better employment,
   Till Ponto and Don can get out,
I'll cultivate rural enjoyment,
   And angle immensely for trout.

Good-night to the Season!—the lobbies,
   Their changes, and rumours of change,
Which startled the rustic Sir Bobbies,
   And made all the Bishops look strange:
The breaches, and battles, and blunders,
   Perform'd by the Commons and Peers;
The Marquis's eloquent thunders,
   The Baronet's eloquent ears:
Denouncings of Papists and treasons,
   Of foreign dominion and oats;
Misrepresentations of reasons,
   And misunderstandings of notes.

Good-night to the Season!—the buildings
  Enough to make Inigo sick;
The paintings, and plasterings, and gildings
  Of stucco, and marble, and brick;
The orders deliciously blended,
  From love of effect, into one;
The club-houses only intended,
  The palaces only begun;
The hell where the fiend, in his glory,
  Sits staring at putty and stones,
And scrambles from story to story,
  To rattle at midnight his bones.

Good-night to the Season!—the dances,
  The fillings of hot little rooms,
The glancings of rapturous glances,
  The fancyings of fancy costumes;
The pleasures which Fashion makes duties,
  The praisings of fiddles and flutes,
The luxury of looking at beauties,
  The tedium of talking to mutes;
The female diplomatists, planners
  Of matches for Laura and Jane,
The ice of her Ladyship's manners,
  The ice of his Lordship's champagne.

Good-night to the Season!—the rages
  Led off by the chiefs of the throng,
The Lady Matilda's new pages,
  The Lady Eliza's new song;
Miss Fennel's macaw, which at Boodle's
  Is held to have something to say;
Mrs Splenetic's musical poodles,
  Which bark 'Batti Batti' all day;
The pony Sir Araby sported,
  As hot and as black as a coal,
And the Lion his mother imported,
  In bearskins and grease, from the Pole.

Good-night to the Season!—the Toso,
  So very majestic and tall;
Miss Ayton, whose singing was so-so,
  And Pasta, divinest of all;

The labour in vain of the Ballet,
  So sadly deficient in stars;
The foreigners thronging the Alley,
  Exhaling the breath of cigars;
The 'loge' where some heiress, how killing,
  Environ'd with Exquisites sits,
The lovely one out of her drilling,
  The silly ones out of their wits.

Good-night to the Season!—the splendour
  That beam'd in the Spanish Bazaar;
Where I purchased—my heart was so tender—
  A card-case,—a pasteboard guitar,—
A bottle of perfume,—a girdle,—
  A lithograph'd Riego full-grown,
Whom Bigotry drew on a hurdle
  That artists might draw him on stone,—
A small panorama of Seville,—
  A trap for demolishing flies,—
A caricature of the Devil,—
  And a look from Miss Sheridan's eyes.

Good-night to the Season!—the flowers
  Of the grand horticultural fête,
When boudoirs were quitted for bowers,
  And the fashion was not to be late;
When all who had money and leisure
  Grew rural o'er ices and wines,
All pleasantly toiling for pleasure,
  All hungrily pining for pines,
And making of beautiful speeches,
  And marring of beautiful shows,
And feeding on delicate peaches,
  And treading on delicate toes.

Good-night to the Season!—another
  Will come with its trifles and toys,
And hurry away, like its brother,
  In sunshine, and odour, and noise.
Will it come with a rose or a briar?
  Will it come with a blessing or curse?
Will its bonnets be lower or higher?
  Will its morals be better or worse?

Riego] instigator of the Spanish Revolution of 1820

Will it find me grown thinner or fatter,
Or fonder of wrong or of right,
Or married,—or buried?—no matter,
Good-night to the Season, Good-night!

JAMES CLARENCE MANGAN

1803–1849

118 *The Woman of Three Cows*

O WOMAN of Three Cows, *agra!* don't let your tongue thus rattle!
O, don't be saucy, don't be stiff, because you may have cattle.
I have seen—and, here's my hand to you, I only say what's true—
A many a one with twice your stock not half so proud as you.

Good luck to you, don't scorn the poor, and don't be their despiser,
For worldly wealth soon melts away, and cheats the very miser,
And Death soon strips the proudest wreath from haughty human brows;
Then don't be stiff, and don't be proud, good Woman of Three Cows!

See where Momonia's heroes lie, proud Owen More's descendants,
'Tis they that won the glorious name, and had the grand attendants!
If *they* were forced to bow to Fate, as every mortal bows,
Can *you* be proud, can *you* be stiff, my Woman of Three Cows!

The brave sons of the Lord of Clare, they left the land to mourning;
Mavrone! for they were banished, with no hope of their returning—
Who knows in what abodes of want those youths were driven to house?
Yet *you* can give yourself these airs, O Woman of Three Cows!

O, think of Donnell of the Ships, the Chief whom nothing daunted—
See how he fell in distant Spain, unchronicled, unchanted!
He sleeps, the great O'Sullivan, where thunder cannot rouse—
Then ask yourself, should *you* be proud, good Woman of Three Cows!

agra] my love   *Mavrone*] my sorrow

O'Ruark, Maguire, those souls of fire, whose names are shrined in story—
Think how their high achievements once made Erin's highest glory—
Yet now their bones lie mouldering under weeds and cypress boughs,
And so, for all your pride, will yours, O Woman of Three Cows!

The O'Carrolls, also, famed when Fame was only for the boldest,
Rest in forgotten sepulchres with Erin's best and oldest;
Yet who so great as they of yore in battle or carouse?
Just think of that, and hide your head, good Woman of Three Cows!

Your neighbour's poor, and you, it seems, are big with vain ideas,
Because, *inagh!* you've got three cows—one more, I see, than *she* has.
That tongue of yours wags more at times than Charity allows,
But if you're strong, be merciful, great Woman of Three Cows!

Now, there you go! You still, of course, keep up your scornful bearing,
And I'm too poor to hinder you; but, by the cloak I'm wearing,
If I had but *four* cows myself, even though you were my spouse,
I'd thwack you well to cure your pride, my Woman of Three Cows!

BENJAMIN HALL KENNEDY

1804–1889

119 *On* Who wrote Icon Basilike? *by Dr Christopher Wordsworth, Master of Trinity*

WHO wrote *Who wrote Icon Basilike?*
I, said the Master of Trinity,
With my small ability,
I wrote *Who wrote Icon Basilike?*

inagh] forsooth

CHARLES LEVER

1806–1872

120

Bad Luck to This Marching

BAD luck to this marching,
Pipeclaying and starching
How neat one must be to be killed by the French!
I'm sick of parading,
Through wet and cold wading,
Or standing all night to be shot in a trench.
To the tune of a fife
They dispose of your life,
You surrender your soul to some illigant lilt;
Now I like 'Garryowen'
When I hear it at home,
But it's not half so sweet when you're going to be kilt.

Then, though up late and early
Our pay comes so rarely,
The devil a farthing we've ever to spare;
They say some disaster
Befell the paymaster;
On my conscience, I think that the money's not there.
And, just think, what a blunder,
They won't let us plunder,
While the convents invite us to rob them, 'tis clear;
Though there isn't a village
But cries, 'Come and pillage!'
Yet we leave all the mutton behind for Mounseer.

Like a sailor that's nigh land,
I long for that island
Where even the kisses we steal if we please;
Where it is no disgrace
If you don't wash your face,
And you've nothing to do but to stand at your ease.
With no sergeant to abuse us,
We fight to amuse us,
Sure it's better beat Christians than kick a baboon;
How I'd dance like a fairy
To see ould Dunleary,
And think twice ere I'd leave it to be a dragoon!

ALFRED, LORD TENNYSON

1809–1892

121 *Northern Farmer: New Style*

I

DOSN'T thou 'ear my 'erse's legs, as they canters awaäy?
Proputty, proputty, proputty—that's what I 'ears 'em saäy.
Proputty, proputty, proputty—Sam, thou's an ass for thy paaïns:
Theer's moor sense i' one o' 'is legs nor in all thy braaïns.

II

Woä—theer's a craw to pluck wi' tha, Sam: yon's parson's 'ouse—
Dosn't thou knaw that a man mun be eäther a man or a mouse?
Time to think on it then; for thou'll be twenty to weeäk.
Proputty, proputty—woä then woä—let ma 'ear mysen speäk.

III

Me an' thy muther, Sammy, 'as beän a-talkin' o' thee;
Thou's beän talkin' to muther, an' she beän a tellin' it me.
Thou'll not marry for munny—thou's sweet upo' parson's lass—
Noä—thou'll marry for luvv—an' we boäth on us thinks tha an ass.

IV

Seeä'd her todaäy goä by—Saäint's daäy—they was ringing the bells.
She's a beauty thou thinks—an' soä is scoors o' gells,
Them as 'as munny an' all—wot's a beauty?—the flower as blaws.
But proputty, proputty sticks, an' proputty, proputty graws.

V

Do'ant be stunt: taäke time: I knaws what maäkes tha sa mad.
Warn't I craäzed fur the lasses mysen when I wur a lad?
But I knaw'd a Quaäker feller as often 'as towd ma this:
'Doänt thou marry for munny, but goä wheer munny is!'

VI

An' I went wheer munny war: an' thy muther coom to 'and,
Wi' lots o' munny laaïd by, an' a nicetish bit o' land.
Maäybe she warn't a beauty:—I niver giv it a thowt—
But warn't she as good to cuddle an' kiss as a lass as 'ant nowt?

to weeäk] this week *stunt*] obstinate

VII

Parson's lass 'ant nowt, an' she weänt 'a nowt when 'e's deäd,
Mun be a guvness, lad, or summut, and addle her breäd:
Why? fur 'e's nobbut a curate, an' weänt niver git hissen clear,
An' 'e maäde the bed as 'e ligs on afoor 'e coom'd to the shere.

VIII

An thin 'e coom'd to the parish wi' lots o' Varsity debt,
Stook to his taaïl they did, an' 'e 'ant got shut on 'em yet.
An' 'e ligs on 'is back i' the grip, wi' noän to lend 'im a shuvv,
Woorse nor a far-welter'd yowe: fur, Sammy, 'e married fur luvv.

IX

Luvv? what's luvv? thou can luvv thy lass an' 'er munny too,
Maakin' 'em goä togither as they've good right to do.
Could'n I luvv thy muther by cause o' 'er munny laaïd by?
Naäy—fur I luvv'd 'er a vast sight moor fur it: reäson why.

X

Ay an' thy muther says thou wants to marry the lass,
Cooms of a gentleman burn: an' we boäth on us thinks tha an ass.
Woä then, proputty, wiltha?—an ass as near as mays nowt—
Woä then, wiltha? dangtha!—the bees is as fell as owt.

XI

Breäk me a bit o' the esh for his 'eäd lad, out o' the fence!
Gentleman burn! what's gentleman burn? is it shillins an' pence?
Proputty, proputty's ivrything 'ere, an', Sammy, I'm blest
If it isn't the saäme oop yonder, fur them as 'as it's the best.

XII

Tis'n them as 'as munny as breäks into 'ouses an' steäls,
Them as 'as coats to their backs an' taäkes their regular meäls.
Noä, but it's them as niver knaws wheer a meäl's to be 'ad.
Taäke my word for it, Sammy, the poor in a loomp is bad.

XIII

Them or thir feythers, tha sees, mun 'a beän a laäzy lot,
Fur work mun 'a gone to the gittin' whiniver munny was got.
Feyther 'ad ammost nowt; leästways 'is munny was 'id.
But 'e tued an' moil'd 'issen deäd, an 'e died a good un, 'e did.

addle] earn *a far-welter'd yowe*] a sheep lying on its back *mays nowt*] makes nothing *the bees is as fell as owt*] the flies are as fierce as anything

XIV

Look thou theer wheer Wrigglesby beck cooms out by the 'ill!
Feyther run oop to the farm, an' I runs oop to the mill;
An' I'll run oop to the brig, an' that thou'll live to see;
And if thou marries a good un I'll leäve the land to thee.

XV

Thim's my noätions, Sammy, wheerby I means to stick;
But if thou marries a bad un, I'll leäve the land to Dick.—
Coom oop, proputty, proputty—that's what I 'ears 'im saäy—
Proputty, proputty, proputty—canter an' canter awaäy.

OLIVER WENDELL HOLMES

1809–1894

122 *Cacoëthes Scribendi*

IF all the trees in all the woods were men,
And each and every blade of grass a pen;
If every leaf on every shrub and tree
Turned to a sheet of foolscap; every sea
Were changed to ink, and all earth's living tribes
Had nothing else to do but act as scribes,
And for ten thousand ages, day and night,
The human race should write, and write, and write,
Till all the pens and paper were used up,
And the huge inkstand was an empty cup,
Still would the scribblers clustered round its brink
Call for more pens, more paper, and more ink.

123 *At The 'Atlantic' Dinner, December 15, 1874*

I SUPPOSE it's myself that you're making allusion to
And bringing the sense of dismay and confusion to.
Of course *some* must speak,—they are always selected to,
But pray what's the reason that I am expected to?

122 *Cacoëthes Scribendi*] an itch for scribbling

123 *'Atlantic'*] the *Atlantic Monthly*, founded in 1857

I'm not fond of wasting my breath as those fellows do
That want to be blowing for ever as bellows do;
Their legs are uneasy, but why will you jog any
That long to stay quiet beneath the mahogany?

Why, why call *me* up with your battery of flatteries?
You say 'He writes poetry,'—that's what the matter is!
'It costs him no trouble—a penful of ink or two
And the poem is done in the time of a wink or two;
As for thoughts—never mind—take the ones that lie uppermost,
And the rhymes used by Milton and Byron and Tupper most;
The lines come so easy! at one end he jingles 'em,
At the other with capital letters he shingles 'em,—
Why, the thing writes itself, and before he's half done with it
He hates to stop writing, he has such good fun with it!'

Ah, that is the way in which simple ones go about
And draw a fine picture of things they don't know about!
We all know a kitten, but come to a catamount
The beast is a stranger when grown up to that amount
(A stranger we rather prefer shouldn't visit us,
A *felis* whose advent is far from felicitous).
The boy who can boast that his trap has just got a mouse
Mustn't draw it and write underneath 'hippopotamus;'
Or say unveraciously, 'This is an elephant'—
Don't think, let me beg, these examples irrelevant—
What they mean is just this—that a thing to be painted well
Should always be something with which we're acquainted well.

You call on your victim for 'things he has plenty of,—
Those copies of verses no doubt at least twenty of;
His desk is crammed full, for he always keeps writing 'em
And reading to friends as his way of delighting 'em!'—
I tell you this writing of verses means business,—
It makes the brain whirl in a vortex of dizziness:
You think they are scrawled in the languor of laziness—
I tell you they're squeezed by a spasm of craziness,
A fit half as bad as the staggering vertigos
That seize a poor fellow and down in the dirt he goes!

And therefore it chimes with the word's etymology
That the sons of Apollo are great on apology,
For the writing of verse is a struggle mysterious
And the gayest of rhymes is a matter that's serious.

For myself, I'm relied on by friends in extremities,
And I don't mind so much if a comfort to them it is;
'Tis a pleasure to please, and the straw that can tickle us
Is a source of enjoyment, though slightly ridiculous.

I am up for a—something—and since I've begun with it,
I must give you a toast now before I have done with it.
Let me pump at my wits as they pumped the Cochituate
That moistened—it may be—the very last bit you ate.

—Success to our publishers, authors, and editors;
To our debtors good luck,—pleasant dreams to our creditors;
May the monthly grow yearly, till all we are groping for
Has reached the fulfilment we're all of us hoping for;
Till the bore through the tunnel—it makes me let off a sigh
To think it may possibly ruin my prophecy—
Has been punned on so often 'twill never provoke again
One mild adolescent to make the old joke again;
Till abstinent, all-go-to-meeting society
Has forgotten the sense of the word inebriety;
Till the work that poor Hannah and Bridget and Phillis do
The humanised, civilised female gorillas do;
Till the roughs, as we call them, grown loving and dutiful,
Shall worship the true and the pure and the beautiful,
And, preying no longer as tiger and vulture do,
All read the 'Atlantic' as persons of culture do!

WILLIAM MAKEPEACE THACKERAY

1811–1863

124

Sorrows of Werther

WERTHER had a love for Charlotte
  Such as words could never utter;
Would you know how first he met her?
  She was cutting bread-and-butter.

Charlotte was a married lady,
And a moral man was Werther,
And, for all the wealth of Indies,
Would do nothing for to hurt her.

So he sighed and pined and ogled,
And his passion boiled and bubbled,
Till he blew his silly brains out,
And no more was by it troubled.

Charlotte, having seen his body
Borne before her on a shutter,
Like a well-conducted person,
Went on cutting bread-and-butter.

125 *The Speculators*

THE night was stormy and dark, The town was shut up in sleep: Only those were abroad who were out on a lark, Or those who'd no beds to keep.

I pass'd through the lonely street, The wind did sing and blow; I could hear the policeman's feet Clapping to and fro.

There stood a potato-man In the midst of all the wet; He stood with his 'tato-can In the lonely Haymarket.

Two gents of dismal mien, And dank and greasy rags Came out of a shop for gin, Swaggering over the flags:

Swaggering over the stones, These shabby bucks did walk; And I went and followed those seedy ones, And listened to their talk.

Was I sober or awake? Could I believe my ears? Those dismal beggars spake Of nothing but railroad shares.

I wondered more and more: Says one—'Good friend of mine, How many shares have you wrote for? In the Diddlesex Junction line?'

'I wrote for twenty,' says Jim, 'But they wouldn't give me one;' His comrade straight rebuked him For the folly he had done:

'O Jim, you are unawares Of the ways of this bad town; *I* always write for five hundred shares, And *then* they put me down.'

'And yet you got no share,' Says Jim, 'for all your boast;' 'I *would* have wrote,' says Jack, 'but where Was the penny to pay the post?'

'I lost, for I couldn't pay That first instalment up; But here's taters smoking hot—I say Let's stop my boy and sup.'

And at this simple feast The while they did regale, I drew each ragged capitalist Down on my left thumb-nail.

Their talk did me perplex, All night I tumbled and tost, And thought of railroad specs., And how money was won and lost.

'Bless railroads everywhere,' I said, 'and the world's advance; Bless every railroad share In Italy, Ireland, France; For never a beggar need now despair, And every rogue has a chance.'

126 *Dear Jack*

DEAR Jack, this white mug that with Guinness I fill,
And drink to the health of sweet Nan of the Hill,
Was once Tommy Tosspot's, as jovial a sot,
As e'er drew a spigot, or drain'd a full pot—
In drinking, all round 'twas his joy to surpass,
And with all merry tipplers he swigg'd off his glass.

One morning in summer, while seated so snug,
In the porch of his garden, discussing his jug,
Stern Death, on a sudden, to Tom did appear,
And said, 'Honest Thomas, come take your last bier;'
We kneaded his clay in the shape of this can,
From which let us drink to the health of my Nan.

ROBERT BROWNING

1812–1889

127 from *The Flight of the Duchess*

AND were I not, as a man may say, cautious
How I trench, more than needs, on the nauseous,

I could favour you with sundry touches
Of the paint-smutches with which the Duchess
Heightened the mellowness of her cheek's yellowness
(To get on faster) until at last her
Cheek grew to be one master-plaster
Of mucus and fucus from mere use of ceruse:
In short, she grew from scalp to udder
Just the object to make you shudder.

128 *Rhyme for a Child Viewing a Naked Venus in a Painting*

HE gazed and gazed and gazed and gazed,
Amazed, amazed, amazed, amazed.

129 *Dialogue between Father and Daughter*

F. THEN, what do you say to the poem of Mizpah?
D. An out-and-out masterpiece—that's what it is, Pa!

EDWARD LEAR

1812–1888

130 *Limericks, I*

(i)

THERE was an Old Man in a boat,
Who said, 'I'm afloat! I'm afloat!'
When they said, 'No! you ain't!' he was ready to faint,
That unhappy Old Man in a boat.

(ii)

THERE was an old Lady of Winchelsea,
Who said, 'If you needle or pin shall see,
On the floor of my room, sweep it up with the broom!'
—That exhaustive old Lady of Winchelsea!

(iii)

THERE was an Old Person of Basing,
Whose presence of mind was amazing;
He purchased a steed, which he rode at full speed,
And escaped from the people of Basing.

(iv)

THERE was an Old Person of Gretna,
Who rushed down the crater of Etna;
When they said, 'Is it hot?' He replied, 'No, it's not!'
That mendacious Old Person of Gretna.

(v)

THERE was a Young Lady of Parma,
Whose conduct grew calmer and calmer;
When they said, 'Are you dumb?' she merely said, 'Hum!'
That provoking Young Lady of Parma.

131 *The Owl and The Pussy-Cat*

I

THE Owl and the Pussy-cat went to sea
  In a beautiful pea-green boat,
They took some honey, and plenty of money,
  Wrapped up in a five-pound note.
The Owl looked up to the stars above,
  And sang to a small guitar,
'O lovely Pussy! O Pussy, my love,
  What a beautiful Pussy you are,
      You are,
      You are!
  What a beautiful Pussy you are!'

II

Pussy said to the Owl, 'You elegant fowl!
  How charmingly sweet you sing!
O let us be married! too long we have tarried:
  But what shall we do for a ring?'
They sailed away, for a year and a day,

To the land where the Bong-tree grows
And there in a wood a Piggy-wig stood
With a ring at the end of his nose,
His nose,
His nose,
With a ring at the end of his nose.

III

'Dear Pig, are you willing to sell for one shilling
Your ring?' Said the Piggy, 'I will.'
So they took it away, and were married next day
By the Turkey who lives on the hill.
They dined on mince, and slices of quince,
Which they ate with a runcible spoon;
And hand in hand, on the edge of the sand,
They danced by the light of the moon,
The moon,
The moon,
They danced by the light of the moon.

132 *'Our Mother Was the Pussy-Cat'*

(An incomplete draft, found among Lear's papers)

OUR mother was the Pussy-cat, our father was the Owl,
And so we're partly little beasts and partly little fowl,
The brothers of our family have feathers and they hoot,
While all the sisters dress in fur and have long tails to boot.
We all believe that little mice,
For food are singularly nice.
Our mother died long years ago. She was a lovely cat
Her tail was 5 feet long, and grey with stripes, but what of that?
In Sila forest on the East of far Calabria's shore
She tumbled from a lofty tree—none ever saw her more.
Our owly father long was ill from sorrow and surprise,
But with the feathers of his tail he wiped his weeping eyes.
And in the hollow of a tree in Sila's inmost maze
We made a happy home and there we pass our obvious days.

From Reggian Cosenza many owls about us flit
And bring us worldly news for which we do not care a bit.
We watch the sun each morning rise, beyond Tarento's strait;
We go out ———— before it gets too late;

And when the evening shades begin to lengthen from the trees
——————— as sure as bees is bees.
We wander up and down the shore ———————
Or tumble over head and heels, but never, never more
Can see the far Gromboolian plains ———————
Or weep as we could once have wept o'er many a vanished scene:
This is the way our father moans—he is so very green.

Our father still preserves his voice, and when he sees a star
He often sings ——————— to that original guitar.
———————————————
———————————————

The pot in which our parents took the honey in their boat,
But all the money has been spent, beside the £5 note.
The owls who come and bring us news are often——
Because we take no interest in poltix of the day.)

133 *Limericks, II*

(i)

THERE was a young person whose history,
Was always considered a mystery;
She sate in a ditch, although no one knew which,
And composed a small treatise on history.

(ii)

THERE was an Old Lady of Chertsey,
Who made a remarkable curtsey;
She twirled round and round, till she sunk underground,
Which distressed all the people of Chertsey.

(iii)

THERE was an old man of Toulouse
Who purchased a new pair of shoes;
When they asked, 'Are they pleasant?'—He said, 'Not at present!'
That turbid old man of Toulouse.

(iv)

THERE was a Young Lady of Portugal,
Whose ideas were excessively nautical:
She climbed up a tree, to examine the sea,
But declared she would never leave Portugal.

(v)

THERE was an old man whose despair
Induced him to purchase a hare:
Whereon one fine day, he rode wholly away,
Which partly assuaged his despair.

134 *Incidents in the Life of my Uncle Arly*

I

O MY aged Uncle Arly!
Sitting on a heap of Barley
    Thro' the silent hours of night,—
Close beside a leafy thicket:—
On his nose there was a Cricket,—
In his hat a Railway-Ticket;—
    (But his shoes were far too tight.)

II

Long ago, in youth, he squander'd
All his goods away, and wander'd
    To the Tiniskoop-hills afar.
There on golden sunsets blazing,
Every evening found him gazing,—
Singing,—'Orb! you're quite amazing!
    'How I wonder what you are!'

III

Like the ancient Medes and Persians,
Always by his own exertions
    He subsisted on those hills;—
Whiles,—by teaching children spelling,—
Or at times by merely yelling,—
Or at intervals by selling
    Propter's Nicodemus Pills.

IV

Later, in his morning rambles
He perceived the moving brambles—
    Something square and white disclose;—
'Twas a First-class Railway-Ticket;
But, on stooping down to pick it
Off the ground,—a pea-green Cricket
    Settled on my uncle's Nose.

V

Never—never more,—oh! never,
Did that Cricket leave him ever,—
    Dawn or evening, day or night;—
Clinging as a constant treasure,—
Chirping with a cheerious measure,—
Wholly to my uncle's pleasure,—
    (Though his shoes were far too tight.)

VI

So for three-and-forty winters,
Till his shoes were worn to splinters,
    All those hills he wander'd o'er,—
Sometimes silent;—sometimes yelling;—
Till he came to Borley-Melling,
Near his old ancestral dwelling;—
    (But his shoes were far too tight.)

VII

On a little heap of Barley
Died my aged uncle Arly,
    And they buried him one night;—
Close beside the leafy thicket;—
There,—his hat and Railway-Ticket;—
There,—his ever-faithful Cricket;—
    (But his shoes were far too tight.)

'BON GAULTIER'

William Aytoun (1813–1865) and Sir Theodore Martin (1816–1909)

135 from *The American's Apostrophe to Boz*

WE received thee warmly—kindly—though we knew thou wert a quiz,
Partly for thyself it may be, chiefly for the sake of Phiz!
Much we bore, and much we suffered, listening to remorseless spells
Of that Smike's unceasing drivellings, and these everlasting Nells.

When you talked of babes and sunshine, fields, and all that sort of thing,
Each Columbian inly chuckled, as he slowly sucked his sling;
And though all our sleeves were bursting, from the many hundreds near
Not one single scornful titter rose on thy complacent ear.

. . . .

Did we spare our brandy-cocktails, stint thee of our whisky-grogs?
Half the juleps that we gave thee would have floored a Newman Noggs;
And thou took'st them in so kindly, little was there then to blame,
To thy parched and panting palate sweet as mother's milk they came.
Did the hams of old Virginny find no favour in thine eyes?
Came no soft compunction o'er thee at the thought of pumpkin pies?
Could not all our chicken fixings into silence fix thy scorn?
Did not all our cakes rebuke thee, Johnny, waffle, dander, corn?
Could not all our care and coddling teach thee how to draw it mild?
Well, no matter, we deserve it. Serves us right! We spoilt the child!
You, forsooth, must come crusading, boring us with broadest hints
Of your own peculiar losses by American reprints.
Such an impudent remonstrance never in our face was flung;
Lever stands it, so does Ainsworth; *you*, I guess, may hold your tongue.
Down our throats you'd cram your projects, thick and hard as pickled salmon.
That, I s'pose, you call free trading,—I pronounce it utter gammon.
No, my lad, a 'cuter vision than your own might soon have seen
That a true Columbian ogle carries little that is green;
That we never will surrender useful privateering rights,
Stoutly won at glorious Bunker's Hill, and other famous fights;
That we keep our native dollars for our native scribbling gents,
And on British manufacture only waste our straggling cents;
Quite enough we pay, I reckon, when we stump of these a few
For the voyages and travels of a freshman such as you.

136 *The Royal Banquet*

THE Queen she kept high festival in Windsor's lordly hall,
And round her sat the gartered knights, and ermined nobles all;
There drank the valiant Wellington, there fed the wary Peel,
And at the bottom of the board Prince Albert carved the veal.

'What, pantler, ho! remove the cloth! Ho! cellarer, the wine,
And bid the royal nurse bring in the hope of Brunswick's line!'
Then rose with one tumultuous shout the band of British peers,
'God bless her sacred Majesty! Let's see the little dears!'

Now by Saint George, our patron saint, 'twas a touching sight to see
That iron warrior gently place the Princess on his knee;
To hear him hush her infant fears, and teach her how to gape
With rosy mouth expectant for the raisin and the grape!

They passed the wine, the sparkling wine—they filled the goblets up;
Even Brougham, the cynic anchorite, smiled blandly on the cup;
And Lyndhurst, with a noble thirst, that nothing could appease,
Proposed the immortal memory of King William on his knees.

'What want we here, my gracious liege,' cried gay Lord Aberdeen,
'Save gladsome song and minstrelsy to flow our cups between?
I ask not now for Goulburn's voice or Knatchbull's warbling lay,
But where's the Poet Laureate to grace our board to-day?'

Loud laughed the Knight of Netherby, and scornfully he cried,
'Or art thou mad with wine, Lord Earl, or art thyself beside?
Eight hundred Bedlam bards have claimed the Laureate's vacant crown,
And now like frantic Bacchanals run wild through London town!'

'Now glory to our gracious Queen!' a voice was heard to cry,
And dark Macaulay stood before them all with frenzied eye;
'Now glory to our gracious Queen, and all her glorious race,
A boon, a boon, my sovran liege! Give me the Laureate's place!

''Twas I that sang the might of Rome, the glories of Navarre;
And who could swell the fame so well of Britain's Isles afar?
The hero of a hundred fights—' Then Wellington up sprung,
'Ho, silence in the ranks, I say! Sit down and hold your tongue!

'By heaven, thou shalt not twist my name into a jingling lay,
Or mimic in thy puny song the thunders of Assaye!
'Tis hard that for thy lust of place in peace we cannot dine.
Nurse, take her Royal Highness, here! Sir Robert, pass the wine!'

'No Laureate need we at our board!' then spoke the Lord of Vaux;
'Here's many a voice to charm the ear with minstrel song, I know.
Even I myself—' Then rose the cry—'A song, a song from Brougham!'
He sang,—and straightway found himself alone within the room.

ARTHUR HUGH CLOUGH

1819–1861

137

Spectator ab extra

I

As I sat at the Café I said to myself,
They may talk as they please about what they call pelf,
They may sneer as they like about eating and drinking,
But help it I cannot, I cannot help thinking
  How pleasant it is to have money, heigh-ho!
  How pleasant it is to have money.

I sit at my table *en grand seigneur*,
And when I have done, throw a crust to the poor;
Not only the pleasure itself of good living,
But also the pleasure of now and then giving:
  So pleasant it is to have money, heigh-ho!
  So pleasant it is to have money.

They may talk as they please about what they call pelf,
And how one ought never to think of one's self,
How pleasures of thought surpass eating and drinking—
My pleasure of thought is the pleasure of thinking
  How pleasant it is to have money, heigh-ho!
  How pleasant it is to have money.

II

Le Dîner

Come along, 'tis the time, ten or more minutes past,
And he who came first had to wait for the last;
The oysters ere this had been in and been out;
Whilst I have been sitting and thinking about
  How pleasant it is to have money, heigh-ho!
  How pleasant it is to have money.

A clear soup with eggs; *voilà tout*; of the fish
The *filets de sole* are a moderate dish
A la Orly, but you're for red mullet, you say:
By the gods of good fare, who can question to-day
  How pleasant it is to have money, heigh-ho!
  How pleasant it is to have money.

After oysters, sauterne; then sherry; champagne,
Ere one bottle goes, comes another again;
Fly up, thou bold cork, to the ceiling above,
And tell to our ears in the sound that they love
  How pleasant it is to have money, heigh-ho!
  How pleasant it is to have money.

I've the simplest of palates; absurd it may be,
But I almost could dine on a *poulet-au-riz*,
Fish and soup and omelette and that—but the deuce—
There were to be woodcocks, and not *Charlotte Russe*!
  So pleasant it is to have money, heigh-ho!
  So pleasant it is to have money.

Your Chablis is acid, away with the Hock,
Give me the pure juice of the purple Médoc:
St Peray is exquisite; but, if you please,
Some Burgundy just before tasting the cheese.
  So pleasant it is to have money, heigh-ho!
  So pleasant it is to have money.

As for that, pass the bottle, and d——n the expense,
I've seen it observed by a writer of sense,
That the labouring classes could scarce live a day,
If people like us didn't eat, drink, and pay.
  So useful it is to have money, heigh-ho!
  So useful it is to have money.

One ought to be grateful, I quite apprehend,
Having dinner and supper and plenty to spend,
And so suppose now, while the things go away,
By way of a grace we all stand up and say
  How pleasant it is to have money, heigh-ho!
  How pleasant it is to have money.

III

Parvenant

I cannot but ask, in the park and the streets
When I look at the number of persons one meets,
What e'er in the world the poor devils can do
Whose fathers and mothers can't give them a *sou*.
So needful it is to have money, heigh-ho!
So needful it is to have money.

I ride, and I drive, and I care not a d——n,
The people look up and they ask who I am;
And if I should chance to run over a cad,
I can pay for the damage, if ever so bad.
    So useful it is to have money, heigh-ho!
    So useful it is to have money.

It was but this winter I came up to town,
And already I'm gaining a sort of renown;
Find my way to good houses without much ado,
Am beginning to see the nobility too.
    So useful it is to have money, heigh-ho!
    So useful it is to have money.

O dear what a pity they ever should lose it,
Since they are the people that know how to use it;
So easy, so stately, such manners, such dinners,
And yet, after all, it is we are the winners.
    So needful it is to have money, heigh-ho!
    So needful it is to have money.

It's all very well to be handsome and tall,
Which certainly makes you look well at a ball;
It's all very well to be clever and witty,
But if you are poor, why it's only a pity.
    So needful it is to have money, heigh-ho!
    So needful it is to have money.

There's something undoubtedly in a fine air,
To know how to smile and be able to stare.
High breeding is something, but well-bred or not,
In the end the one question is, what have you got.
    So needful it is to have money, heigh-ho!
    So needful it is to have money.

And the angels in pink and the angels in blue,
In muslins and moires so lovely and new,
What is it they want, and so wish you to guess,
But if you have money, the answer is Yes.
    So needful, they tell you, is money, heigh-ho!
    So needful it is to have money.

JAMES RUSSELL LOWELL

1819–1891

138 *What Mr Robinson Thinks*

GINERAL B. is a sensible man;
He stays to his home an' looks arter his folks;
He draws his furrer ez straight ez he can,
An' into nobody's tater-patch pokes;
But John P.
Robinson, he
Sez he wunt vote for Gineral B.

My! ain't it terrible? Wut shall we do?
We can't never choose him, o' course—that's flat:
Guess we shall hev to come round (don't you?),
An' go in for thunder an' guns, an' all that;
Fer John P.
Robinson, he
Sez he wunt vote for Gineral B.

Gineral C. is a dreffle smart man:
He's been on all sides that give places or pelf;
But consistency still was a part of his plan—
He's been true to *one* party, and that is himself;
So John P.
Robinson, he
Sez he shall vote fer Gineral C.

Gineral C. goes in for the war;
He don't vally principle morn 'n an old cud;
What did God make us raytional creeturs fer,
But glory an' gunpowder, plunder an' blood?
So John P.
Robinson, he
Sez he shall vote fer Gineral C.

We're gettin on nicely up here to our village,
With good old idees o' wut's right an' wut ain't;
We kind o' thought Christ went against war and pillage,
An' that eppyletts worn't the best mark of a saint;

war] the American–Mexican war of 1846–8

But John P.
Robinson, he
Sez this kind o' thing's an exploded idee.

The side of our country must ollers be took,
An' President Pulk, you know, *he* is our country;
An' the angel that writes all our sins in a book,
Puts the *debit* to him, an' to us the *per contry*;
An' John P.
Robinson, he
Sez this is his view o' the thing to a T.

Parson Wilbur he calls all these arguments lies;
Sez they're nothin' on airth but jest *fee, faw, fum*;
An' that all this big talk of our destinies
Is half on it ignorance, an' t' other half rum;
But John P.
Robinson, he
Sez it ain't no such thing; an', of course, so must we.

Parson Wilbur sez *he* never heered in his life
Thet the Apostles rigg'd out in their swallow-tail coats,
An' marched round in front of a drum an' a fife,
To git some on 'em office, an' some on 'em votes;
But John P.
Robinson, he
Sez they didn't know everythin' down in Judee.

Wal, it's a marcy we've gut folks to tell us
The rights an' the wrongs o' these matters, I vow—
God sends country lawyers an' other wise fellers
To drive the world's team wen it gits in a slough;
For John P.
Robinson, he
Sez the world'll go right, ef he hollers out Gee!

139 from *A Fable for Critics*

THERE comes Poe, with his raven, like Barnaby Rudge,
Three-fifths of him genius and two-fifths sheer fudge.

FREDERICK LOCKER-LAMPSON

1821–1895

140

A Terrible Infant

I RECOLLECT a nurse call'd Ann,
    Who carried me about the grass,
And one fine day a fine young man
    Came up, and kiss'd the pretty lass.
She did not make the least objection!
        Thinks I, '*Aha!*
    When I can talk I'll tell Mamma'
    —And that's my earliest recollection.

141

My Life Is a ——

AT Worthing, an exile from Geraldine G——
How aimless, how wretched an Exile is he!
Promenades are not even prunella and leather
To lovers, if lovers can't foot them together.

He flies the parade, by the ocean he stands;
He traces a 'Geraldine G.' on the sands;
Only 'G.!' though her loved patronymic is 'Green,'—
'I will not betray thee, my own Geraldine.'

The fortunes of men have a time and a tide,
And Fate, the old Fury, will not be denied;
That name was, of course, soon wiped out by the sea,—
She jilted the Exile, did Geraldine G.

They meet, but they never have spoken since that;
He hopes she is happy,—he knows she is fat;
She woo'd on the shore, now is wed in the Strand;
And *I*—it was I wrote her name on the sand.

THOROLD ROGERS

1823–1890

142

Two Historians

WHILE ladling butter from alternate tubs,
Stubbs butters Freeman, Freeman butters Stubbs.

C. G. LELAND

1824–1903

143

Hans Breitmann's Barty

HANS BREITMANN gife a barty;
  Dey had biano-blayin',
I felled in lofe mit a Merican frau,
  Her name vas Madilda Yane.
She hat haar as prown ash a pretzel,
  Her eyes vas himmel-plue,
Und vhen dey looket indo mine,
  Dey shplit mine heart in dwo.

Hans Breitmann gife a barty,
  I vent dere you'll pe pound;
I valtzet mit Madilda Yane,
  Und vent shpinnen' round und round.
De pootiest Fraulein in de house,
  She vayed 'pout dwo hoondred pound,
Und efery dime she gife a shoomp
  She make de vindows sound.

Hans Breitmann gife a barty,
  I dells you it cost him dear;
Dey rolled in more ash sefen kecks
  Of foost-rate lager beer.
Und vhenefer dey knocks de shpicket in
  De Deutschers gifes a cheer;
I dinks dot so vine a barty
  Nefer coom to a het dis year.

Hans Breitmann gife a barty;
  Dere all vas Souse and Brouse,
Vhen de sooper comed in, de gompany
  Did make demselfs to house;
Dey ate das Brot and Gensy broost,
  De Bratwurst and Braten vine,
Und vash der Abendessen down
  Mit four parrels of Neckarwein.

Hans Breitmann gife a barty;
  Ve all cot troonk ash bigs.
I poot mine mout' to a parrel of beer,
  Und emptied it oop mit a schwigs;
Und den I gissed Madilda Yane,
  Und she shlog me on de kop,
Und de gompany vighted mit daple-lecks
  Dill de coonshtable made oos shtop.

Hans Breitmann gife a barty—
  Vhere ish dot barty now?
Vhere ish de lofely golden cloud
  Dot float on de moundain's prow?
Vhere ish de himmelstrahlende stern—
  De shtar of de shpirit's light?
All goned afay mit de lager beer—
  Afay in de ewigkeit!

ANONYMOUS

144

Epitaph

(Said to have been once found in Bushey Churchyard, Hertfordshire)

HERE lies a poor woman who always was tired,
For she lived in a place where help wasn't hired,
Her last words on earth were, 'Dear friends, I am going,
Where washing ain't done nor cooking nor sewing,
And everything there is exact to my wishes,
For there they don't eat, there's no washing of dishes,

I'll be where loud anthems will always be ringing
(But having no voice, I'll be out of the singing).
Don't mourn for me now, don't grieve for me never,
For I'm going to do nothing for ever and ever.'

MORTIMER COLLINS

1827–1876

145

If!

If life were never bitter,
  And love were always sweet,
Then who would care to borrow
A moral from to-morrow—
If Thames would always glitter,
  And joy would ne'er retreat,
If life were never bitter,
  And love were always sweet.

If care were not the waiter
  Behind a fellow's chair,
When easy-going sinners
Sit down to Richmond dinners,
And life's swift stream flows straighter—
  By Jove, it would be rare,
If care were not the waiter
  Behind a fellow's chair.

If wit were always radiant,
  And wine were always iced,
And bores were kicked out straightway
Through a convenient gateway;
Then down the year's long gradient
  'Twere sad to be enticed,
If wit were always radiant,
  And wine were always iced.

DANTE GABRIEL ROSSETTI

1828–1882

146

On the Painter, Val Prinsep

THERE is a creator called God,
Whose creations are some of them odd.
  I maintain, and I shall,
  The creation of Val
Reflects little credit on God.

147

On the Poet, Arthur O'Shaughnessy

THERE's an Irishman, Arthur O'Shaughnessy—
On the chessboard of poets a pawn is he;
  Though a bishop or king
  Would be rather the thing
To the fancy of Arthur O'Shaughnessy.

EMILY DICKINSON

1830–1886

148

I'm Nobody! Who Are you?

I'M Nobody! Who are you?
Are you—Nobody—Too?
Then there's a pair of us?
Don't tell! they'd advertise—you know!

How dreary—to be—Somebody!
How public—like a Frog—
To tell one's name—the livelong June—
To an admiring Bog!

C. S. CALVERLEY

1831–1884

149 *A B C*

A is an Angel of blushing eighteen:
B is the Ball where the Angel was seen:
C is her Chaperon, who cheated at cards:
D is the Deuxtemps, with Frank of the Guards:
E is her Eye, killing slowly but surely:
F is the Fan, whence it peeped so demurely:
G is the Glove of superlative kid:
H is the Hand which it spitefully hid:
I is the Ice which the fair one demanded:
J is the Juvenile, that dainty who handed:
K is the Kerchief, a rare work of art:
L is the Lace which composed the chief part:
M is the old Maid who watch'd the chits dance:
N is the Nose she turned up at each glance:
O is the Olga (just then in its prime):
P is the Partner who wouldn't keep time:
Q 's a Quadrille, put instead of the Lancers:
R is the Remonstrances made by the dancers:
S is the Supper, where all went in pairs:
T is the Twaddle they talked on the stairs:
U is the Uncle who 'thought we'd be goin':'
V is the Voice which his niece replied 'No' in:
W is the Waiter, who sat up till eight:
X is his Exit, not rigidly straight:
Y is a Yawning fit caused by the Ball:
Z stands for Zero, or nothing at all.

150 *In the Gloaming*

In the Gloaming to be roaming, where the crested waves are foaming,
And the shy mermaidens combing locks that ripple to their feet;
When the Gloaming is, I never made the ghost of an endeavour
To discover—but whatever were the hour, it would be sweet.

'To their feet,' I say, for Leech's sketch indisputably teaches
That the mermaids of our beaches do not end in ugly tails,

Nor have homes among the corals; but are shod with neat balmorals,
An arrangement no one quarrels with, as many might with scales.

Sweet to roam beneath a shady cliff, of course with some young lady,
Lalage, Neæra, Haidee, or Elaine, or Mary Ann:
Love, you dear delusive dream, you! Very sweet your victims deem you,
When, heard only by the seamew, they talk all the stuff one can.

Sweet to haste, a licensed lover, to Miss Pinkerton the glover,
Having managed to discover what is dear Neæra's 'size':
P'raps to touch that wrist so slender, as your tiny gift you tender,
And to read you're no offender, in those laughing hazel eyes.

Then to hear her call you 'Harry,' when she makes you fetch and carry—
O young men about to marry, what a blessed thing it is!
To be photograph'd—together—cased in pretty Russia leather—
Hear her gravely doubting whether they have spoilt your honest phiz!

Then to bring your plighted fair one first a ring—a rich and rare one—
Next a bracelet, if she'll wear one, and a heap of things beside;
And serenely bending o'er her, to inquire if it would bore her
To say when her own adorer may aspire to call her bride!

Then, the days of courtship over, with your WIFE to start for Dover
Or Dieppe—and live in clover evermore, whate'er befalls:
For I've read in many a novel that, unless they've souls that grovel,
Folks *prefer* in fact a hovel to your dreary marble halls:

To sit, happy married lovers; Phillis trifling with a plover's
Egg, while Corydon uncovers with a grace the Sally Lunn,
Or dissects the lucky pheasant—that, I think, were passing pleasant;
As I sit alone at present, dreaming darkly of a Dun.

151

Flight

O MEMORY! that which I gave thee
To guard in thy garner yestreen—
Little deeming thou e'er could'st behave thee
Thus basely—hath gone from thee clean!
Gone, fled, as ere autumn is ended
The yellow leaves flee from the oak—
I have lost it for ever, my splendid
Original joke.

What was it? I know I was brushing
My hair when the notion occurred:
I know that I felt myself blushing
As I thought, 'How supremely absurd!
How they'll hammer on floor and on table
As its drollery dawns on them—how
They will quote it'—I wish I were able
To quote it just now.

I had thought to lead up conversation
To the subject—it's easily done—
Then let off, as an airy creation
Of the moment, that masterly pun.
Let it off, with a flash like a rocket's;
In the midst of a dazzled conclave,
Where I sat, with my hands in my pockets,
The only one grave.

I had fancied young Titterton's chuckles,
And old Bottleby's hearty guffaws
As he drove at my ribs with his knuckles,
His mode of expressing applause:
While Jean Bottleby—queenly Miss Janet—
Drew her handkerchief hastily out,
In fits at my slyness—what can it
Have all been about?

I know 'twas the happiest, quaintest
Combination of pathos and fun:
But I've got no idea—the faintest—
Of what was the actual pun.
I think it was somehow connected
With something I'd recently read—
Or heard—or perhaps recollected
On going to bed.

What *had* I been reading? The *Standard*:
'Double Bigamy;' 'Speech of the Mayor.'
And later—eh? yes! I meandered
Through some chapters of Vanity Fair,
How it fuses the grave with the festive!
Yet e'en there, there is nothing so fine—
So playfully, subtly suggestive—
As that joke of mine.

Did it hinge upon 'parting asunder?'
No, I don't part my hair with my brush.
Was the point of it 'hair'? Now I wonder!
Stop a bit—I shall think of it—hush!
There's *hare*, a wild animal—Stuff!
It was something a deal more recondite:
Of that I am certain enough;
And of nothing beyond it.

Hair—*locks!* There are probably many
Good things to be said about those.
Give me time—that's the best guess of any—
'Lock' has several meanings, one knows.
Iron locks—*iron-gray locks*—a 'deadlock'—
That would set up an everyday wit:
Then of course there's the obvious 'wedlock;'
But that wasn't it.

No! mine was a joke for the ages;
Full of intricate meaning and pith;
A feast for your scholars and sages—
How it would have rejoiced Sidney Smith!
'Tis such thoughts that ennoble a mortal
And, singling him out from the herd,
Fling wide immortality's portal—
But what was the word?

Ah me! 'tis a bootless endeavour.
As the flight of a bird of the air
Is the flight of a joke—you will never
See the same one again, you may swear.
'Twas my firstborn, and O how I prized it!
My darling, my treasure, my own!
This brain and none other devised it—
And now it has flown.

152

Ballad

THE auld wife sat at her ivied door,
(*Butter and eggs and a pound of cheese*)
A thing she had frequently done before;
And her spectacles lay on her apron'd knees.

The piper he piped on the hill-top high,
(*Butter and eggs and a pound of cheese*)
Till the cow said 'I die,' and the goose ask'd 'Why?'
And the dog said nothing, but search'd for fleas.

The farmer he strode through the square farmyard;
(*Butter and eggs and a pound of cheese*)
His last brew of ale was a trifle hard—
The connexion of which with the plot one sees.

The farmer's daughter hath frank blue eyes;
(*Butter and eggs and a pound of cheese*)
She hears the rooks caw in the windy skies,
As she sits at her lattice and shells her peas.

The farmer's daughter hath ripe red lips;
(*Butter and eggs and a pound of cheese*)
If you try to approach her, away she skips
Over tables and chairs with apparent ease.

The farmer's daughter hath soft brown hair
(*Butter and eggs and a pound of cheese*)
And I met with a ballad, I can't say where,
Which wholly consisted of lines like these.

Part II

She sat with her hands 'neath her dimpled cheeks,
(*Butter and eggs and a pound of cheese*)
And spake not a word. While a lady speaks
There is hope, but she didn't even sneeze.

She sat, with her hands 'neath her crimson cheeks;
(*Butter and eggs and a pound of cheese*)
She gave up mending her father's breeks,
And let the cat roll in her new chemise.

She sat, with her hands 'neath her burning cheeks,
(*Butter and eggs and a pound of cheese*)
And gazed at the piper for thirteen weeks;
Then she follow'd him out o'er the misty leas.

Her sheep follow'd her, as their tails did them.
  (*Butter and eggs and a pound of cheese*)
And this song is consider'd a perfect gem,
  And as to the meaning, it's what you please.

LEWIS CARROLL

1832–1898

153

A Visitor

(Written at the age of 13)

WELL, *if* you must know all the facts, I was merely reading a pamphlet
When what should I hear at the door but a knock as soft as a Zephyr.
I listened and heard it again, so, as loud as I possibly *could* call,
I shouted, 'Don't stand waiting there, come in, let me know who you are, sir!'
Mild he entered the room, with his hat in his hand and his gloves off,
And a meek gentle bow he performed, while my anger was rapidly rising,
'Who *are* you?' I angrily cried, and with hand on his heart as he bowed low,
In the gentlest of tones he replied, 'Your servant, Sir Pokurranshuvvle.'
Didn't I just ring the bell, 'Here, Tom, Dick, George, Andrew!' I bawled out,
'Come here! show this stranger the door!' My summons they heard and they did it:
Soon to the door was he guided; once more he turned to me and bowed low,
And so, with his hand on his heart, with all possible meekness departed.

154

A Long Tale

FURY said to
a mouse, That
he met in the
house, 'Let
us both go
to law: *I*
will prose-
cute *you.*—
Come, I'll
take no de-
nial: We
must have
the trial;
For really
this morn-
ing I've
nothing
to do.'
Said the
mouse to
the cur,
'Such a
trial, dear
sir, With
no jury
or judge,
would
be wast-
ing our
breath.'
'I'll be
judge,
I'll be
jury,'
said
cun-
ning
old
Fury:
'I'll
try
the
whole
cause,
and
con-
demn
you to
death.'

155

Father William

'YOU are old, Father William,' the young man said,
  'And your hair has become very white;
And yet you incessantly stand on your head—
  Do you think, at your age, it is right?'

'In my youth,' Father William replied to his son,
  'I feared it might injure the brain;
But, now that I'm perfectly sure I have none,
  Why, I do it again and again.'

'You are old,' said the youth, 'as I mentioned before,
  And have grown most uncommonly fat;
Yet you turned a back-somersault in at the door—
  Pray, what is the reason of that?'

'In my youth,' said the sage, as he shook his grey locks,
  'I kept all my limbs very supple
By the use of this ointment—one shilling the box—
  Allow me to sell you a couple?'

'You are old,' said the youth, 'and your jaws are too weak
  For anything tougher than suet;
Yet you finished the goose, with the bones and the beak—
  Pray, how did you manage to do it?'

'In my youth,' said his father, 'I took to the law,
  And argued each case with my wife;
And the muscular strength, which it gave to my jaw,
  Has lasted the rest of my life.'

'You are old,' said the youth, 'one would hardly suppose
  That your eye was as steady as ever;
Yet you balanced an eel on the end of your nose—
  What made you so awfully clever?'

'I have answered three questions, and that is enough,'
  Said his father; 'don't give yourself airs!
Do you think I can listen all day to such stuff?
  Be off, or I'll kick you down stairs!'

156 *'Tis the Voice of the Lobster*

I

''Tis the voice of the Lobster: I heard him declare
"You have baked me too brown, I must sugar my hair."
As a duck with its eyelids, so he with his nose
Trims his belt and his buttons, and turns out his toes.
When the sands are all dry, he is gay as a lark,
And will talk in contemptuous tones of the Shark:
But, when the tide rises and sharks are around,
His voice has a timid and tremulous sound.'

II

I passed by his garden, and marked, with one eye,
How the Owl and the Panther were sharing a pie:
The Panther took pie-crust, and gravy, and meat,
While the Owl had the dish as its share of the treat.
When the pie was all finished, the Owl, as a boon,
Was kindly permitted to pocket the spoon:
While the Panther received knife and fork with a growl.
And concluded the banquet by—

157 *Humpty Dumpty's Song*

In winter, when the fields are white,
I sing this song for your delight—

In spring, when woods are getting green,
I'll try and tell you what I mean.

In summer, when the days are long,
Perhaps you'll understand the song:

In autumn, when the leaves are brown,
Take pen and ink, and write it down.

I sent a message to the fish:
I told them 'This is what I wish.'

The little fishes of the sea,
They sent an answer back to me.

The little fishes' answer was
'We cannot do it, Sir, because—'

I sent to them again to say
'It will be better to obey.'

The fishes answered with a grin,
'Why, what a temper you are in!'

I told them once, I told them twice:
They would not listen to advice.

I took a kettle large and new,
Fit for the deed I had to do.

My heart went hop, my heart went thump;
I filled the kettle at the pump.

Then someone came to me and said
'The little fishes are in bed.'

I said to him, I said it plain,
'Then you must wake them up again.'

I said it very loud and clear;
I went and shouted in his ear.

But he was very stiff and proud;
He said 'You needn't shout so loud!'

And he was very proud and stiff;
He said 'I'd go and wake them, if—'

I took a corkscrew from the shelf:
I went to wake them up myself.

And when I found the door was locked,
I pulled and pushed and kicked and knocked.

And when I found the door was shut,
I tried to turn the handle, but—

158 from *The Hunting of the Snark*

Fit the Second

The Bellman's Speech

THE Bellman himself they all praised to the skies—
  Such a carriage, such ease and such grace!
Such solemnity, too! One could see he was wise,
  The moment one looked in his face!

He had bought a large map representing the sea,
  Without the least vestige of land:
And the crew were much pleased when they found it to be
  A map they could all understand.

'What's the good of Mercator's North Poles and Equators,
  Tropics, Zones, and Meridian Lines?'
So the Bellman would cry: and the crew would reply,
  'They are merely conventional signs!

'Other maps are such shapes, with their islands and capes!
  But we've got our brave Captain to thank'
(So the crew would protest) 'that he's bought us the best—
  A perfect and absolute blank!'

This was charming, no doubt: but they shortly found out
  That the Captain they trusted so well
Had only one notion for crossing the ocean,
  And that was to tingle his bell.

He was thoughtful and grave—but the orders he gave
  Were enough to bewilder a crew.
When he cried, 'Steer to starboard, but keep her head larboard!'
  What on earth was the helmsman to do?

Then the bowsprit got mixed with the rudder sometimes:
  A thing, as the Bellman remarked,
That frequently happens in tropical climes,
  When a vessel is, so to speak, 'snarked.'

But the principal failing occurred in the sailing,
  And the Bellman, perplexed and distressed,
Said he *had* hoped, at least, when the wind blew due East
  That the ship would *not* travel due West!

But the danger was past—they had landed at last,
With their boxes, portmanteaus, and bags:
Yet at first sight the crew were not pleased with the view,
Which consisted of chasms and crags.

The Bellman perceived that their spirits were low,
And repeated in musical tone
Some jokes he had kept for a season of woe—
But the crew would do nothing but groan.

He served out some grog with a liberal hand,
And bade them sit down on the beach:
And they could not but own that their Captain looked grand,
As he stood and delivered his speech.

'Friends, Romans, and countrymen, lend me your ears!'
(They were all of them fond of quotations:
So they drank to his health, and they gave him three cheers,
While he served out additional rations.)

'We have sailed many months, we have sailed many weeks
(Four weeks to the month you may mark),
But never as yet ('tis your Captain who speaks)
Have we caught the least glimpse of a Snark!

'We have sailed many weeks, we have sailed many days
(Seven days to the week I allow),
But a Snark, on the which we might lovingly gaze,
We have never beheld till now!

'Come, listen, my men, while I tell you again
The five unmistakable marks
By which you may know, wheresoever you go,
The warranted genuine Snarks.

'Let us take them in order. The first is the taste,
Which is meagre and hollow, but crisp:
Like a coat that is rather too tight in the waist,
With a flavour of Will-o'-the-wisp.

'Its habit of getting up late you'll agree
That it carries too far, when I say
That it frequently breakfasts at five-o'clock tea,
And dines on the following day.

'The third is its slowness in taking a jest,
  Should you happen to venture on one,
It will sigh like a thing that is deeply distressed:
  And it always looks grave at a pun.

'The fourth is its fondness for bathing-machines,
  Which it constantly carries about,
And believes that they add to the beauty of scenes—
  A sentiment open to doubt.

'The fifth is ambition. It next will be right
  To describe each particular batch:
Distinguishing those that have feathers, and bite,
  From those that have whiskers, and scratch.

'For, although common Snarks do no manner of harm,
  Yet, I feel it my duty to say,
Some are Boojums—' The Bellman broke off in alarm,
  For the Baker had fainted away.

HARRY CLIFTON

1832–1872

159 *Polly Perkins*

I AM a broken-hearted milkman, in grief I'm arrayed,
Through keeping of the company of a young servant maid,
Who lived on board and wages the house to keep clean
In a gentleman's family near Paddington Green.

Chorus:
    She was as beautiful as a butterfly
      And as proud as a Queen
    Was pretty little Polly Perkins of
      Paddington Green.

She'd an ankle like an antelope and a step like a deer,
A voice like a blackbird, so mellow and clear,
Her hair hung in ringlets so beautiful and long,
I thought that she loved me but I found I was wrong.

When I'd rattle in a morning and cry 'milk below',
At the sound of my milk-cans her face she would show
With a smile upon her countenance and a laugh in her eye,
If I thought she'd have loved me, I'd have laid down to die.

When I asked her to marry me she said 'Oh! what stuff',
And told me to 'drop it, for she had quite enough
Of my nonsense'—at the same time I'd been very kind,
But to marry a milkman she didn't feel inclined.

'Oh, the man that has me must have silver and gold,
A chariot to ride in and be handsome and bold,
His hair must be curly as any watch spring,
And his whiskers as big as a brush for clothing.'

The words that she uttered went straight through my heart,
I sobbed, I sighed, and straight did depart;
With a tear on my eyelid as big as a bean,
Bidding good-bye to Polly and Paddington Green.

In six months she married,—this hard-hearted girl,—
But it was not a Wi-count, and it was not a Nearl,
It was not a 'Baronite', but a shade or two wuss,
It was a bow-legged conductor of a twopenny bus.

GEORGE A. STRONG

1832–1912

160 *The Modern Hiawatha*

He killed the noble Mudjokivis,
With the skin he made him mittens,
Made them with the fur side inside,
Made them with the skin side outside,
He, to get the warm side inside,
Put the inside skin side outside:
He, to get the cold side outside,
Put the warm side fur side inside:
That's why he put the fur side inside,
Why he put the skin side outside,
Why he turned them inside outside.

GEORGE DU MAURIER

1834–1896

161

Vers Nonsensiques

(i)

UN Marin naufragé (de Doncastre)
Pour prière, au milieu du désastre,
    Répétait à genoux
    Ces mots simples et doux:—
'Scintillez, scintillez, petit astre!'

(ii)

A POTSDAM, les totaux absteneurs,
Comme tant d'autres titotalleurs,
    Sont gloutons, omnivores,
    Nasorubicolores,
Grands manchons, et terribles duffeurs.

(iii)

IL existe une Espinstère à Tours,
Un peu vite, et qui porte toujours
    Un ulsteur peau-de-phoque,
    Un chapeau bilicoque,
Et des nicrebocqueurs en velours.

(iv)

A COLOGNE est un maître d'hôtel
Hors du centre du ventre duquel
    Se projette une sorte
    De tiroir qui supporte
La moutarde, et le poivre, et le sel.

(v)

'CASSEZ-VOUS, cassez-vous, cassez-vous,
O mer, sur vos froids gris cailloux!'
    Ainsi traduisait Laure
    Au profit d'Isidore,
(Beau jeune homme, et son futur époux).

W. S. GILBERT

1836–1911

162

There Lived a King

THERE lived a King, as I've been told,
In the wonder-working days of old,
When hearts were twice as good as gold,
    And twenty times as mellow.
Good-temper triumphed in his face,
And in his heart he found a place
For all the erring human race
    And every wretched fellow.
When he had Rhenish wine to drink
It made him very sad to think
That some, at junket or at jink,
    Must be content with toddy.
He wished all men as rich as he
(And he was rich as rich could be),
So to the top of every tree
    Promoted everybody.

Lord Chancellors were cheap as sprats,
And Bishops in their shovel hats
Were plentiful as tabby cats—
    In point of fact, too many.
Ambassadors cropped up like hay,
Prime Ministers and such as they
Grew like asparagus in May,
    And Dukes were three a penny.
On every side Field Marshals gleamed,
Small beer were Lords Lieutenant deemed,
With Admirals the ocean teemed
    All round his wide dominions.
And Party Leaders you might meet
In twos and threes in every street,
Maintaining, with no little heat,
    Their various opinions.

That King, although no one denies
His heart was of abnormal size,
Yet he'd have acted otherwise
    If he had been acuter.

The end is easily foretold,
When every blessed thing you hold
Is made of silver, or of gold,
  You long for simple pewter.
When you have nothing else to wear
But cloth of gold and satins rare,
For cloth of gold you cease to care—
  Up goes the price of shoddy.
In short, whoever you may be,
To this conclusion you'll agree,
When every one is somebodee,
  Then no one's anybody!

163

Ferdinando and Elvira,
or The Gentle Pieman

Part I

AT a pleasant evening party I had taken down to supper
One whom I will call ELVIRA, and we talked of love and TUPPER,

MR TUPPER and the poets, very lightly with them dealing,
For I've always been distinguished for a strong poetic feeling.

Then we let off paper crackers, each of which contained a motto,
And she listened while I read them, till her mother told her not to.

Then she whispered, 'To the ball-room we had better, dear, be walking;
If we stop down here much longer, really people will be talking.'

There were noblemen in coronets, and military cousins,
There were captains by the hundred, there were baronets by dozens.

Yet she heeded not their offers, but dismissed them with a blessing;
Then she let down all her back hair which had taken long in dressing.

Then she had convulsive sobbings in her agitated throttle,
Then she wiped her pretty eyes and smelt her pretty smelling-bottle.

So I whispered, 'Dear ELVIRA, say—what can the matter be with you?
Does anything you've eaten, darling POPSY, disagree with you?'

But spite of all I said, her sobs grew more and more distressing,
And she tore her pretty back hair, which had taken long in dressing.

Then she gazed upon the carpet, at the ceiling then above me,
And she whispered, 'FERDINANDO, do you really, *really* love me?'

'Love you?' said I, then I sighed, and then I gazed upon her sweetly—
For I think I do this sort of thing particularly neatly—

'Send me to the Arctic regions, or illimitable azure,
On a scientific goose-chase, with my COXWELL or my GLAISHER!

'Tell me whither I may hie me, tell me, dear one that I *may* know—
Is it up the highest Andes? down a horrible volcano?'

But she said, 'It isn't polar bears, or hot volcanic grottoes,
Only find out who it is that writes those lovely cracker mottoes!'

Part II

'Tell me, HENRY WADSWORTH, ALFRED, POET CLOSE, or MISTER TUPPER,
Do you write the bonbon mottoes my ELVIRA pulls at supper?'

But HENRY WADSWORTH smiled, and said he had not had that honour;
And ALFRED, too, disclaimed the words that told so much upon her.

'MISTER MARTIN TUPPER, POET CLOSE, I beg of you inform us';
But my question seemed to throw them both into a rage enormous.

MISTER CLOSE expressed a wish that he could only get anigh to me.
And MISTER MARTIN TUPPER sent the following reply to me:—

'A fool is bent upon a twig, but wise men dread a bandit.'
Which I think must have been clever, for I didn't understand it.

Seven weary years I wandered—Patagonia, China, Norway,
Till at last I sank exhausted at a pastrycook his doorway.

There were fuchsias and geraniums, and daffodils and myrtle,
So I entered, and I ordered half a basin of mock turtle.

He was plump and he was chubby, he was smooth and he was rosy,
And his little wife was pretty, and particularly cozy.

And he chirped and sang, and skipped about, and laughed with laughter
hearty—
He was wonderfully active for so very stout a party.

And I said, 'Oh, gentle pieman, why so very, very merry?
Is it purity of conscience, or your one-and-seven sherry?'

But he answered, 'I'm so happy—no profession could be dearer—
If I am not humming "Tra! la! la!" I'm singing "Tirer, lirer!"

'First I go and make the patties, and the puddings and the jellies,
Then I make a sugar birdcage, which upon a table swell is;

'Then I polish all the silver, which a supper-table lacquers;
Then I write the pretty mottoes which you find inside the crackers'—

'Found at last!' I madly shouted. 'Gentle pieman, you astound me!'
Then I waved the turtle soup enthusiastically round me.

And I shouted and I danced until he'd quite a crowd around him—
And I rushed away, exclaiming, 'I have found him! I have found him!'

And I heard the gentle pieman in the road behind me trilling,
'"Tira! lira!" stop him, stop him! "Tra! la! la!" the soup's a shilling!'

But until I reached Elvira's home, I never, never waited,
And Elvira to her Ferdinand's irrevocably mated!

164 *Captain Reece*

Of all the ships upon the blue
No ship contained a better crew
Than that of worthy Captain Reece,
Commanding of *The Mantelpiece*.

He was adored by all his men,
For worthy Captain Reece, RN,
Did all that lay within him to
Promote the comfort of his crew.

If ever they were dull or sad,
Their captain danced to them like mad,
Or told, to make the time pass by,
Droll legends of his infancy.

A feather bed had every man,
Warm slippers and hot-water can,
Brown windsor from the captain's store,
A valet, too, to every four.

Did they with thirst in summer burn?
Lo, seltzogenes at every turn,
And on all very sultry days
Cream ices handed round on trays.

Then currant wine and ginger pops
Stood handily on all the 'tops';
And, also, with amusement rife,
A 'Zoetrope, or Wheel of Life.'

New volumes came across the sea
From MISTER MUDIE's libraree;
The Times and *Saturday Review*
Beguiled the leisure of the crew.

Kind-hearted CAPTAIN REECE, RN,
Was quite devoted to his men;
In point of fact, good CAPTAIN REECE
Beatified *The Mantelpiece*.

One summer eve, at half-past ten,
He said (addressing all his men):
'Come, tell me, please, what I can do
To please and gratify my crew?

'By any reasonable plan
I'll make you happy, if I can;
My own convenience count as *nil*;
It is my duty, and I will.'

Then up and answered WILLIAM LEE
(The kindly captain's coxwain he,
A nervous, shy, low-spoken man),
He cleared his throat and thus began:

'You have a daughter, CAPTAIN REECE,
Ten female cousins and a niece,
A ma, if what I'm told is true,
Six sisters, and an aunt or two.

'Now, somehow, sir, it seems to me,
More friendly-like we all should be
If you united of 'em to
Unmarried members of the crew.

'If you'd ameliorate our life,
Let each select from them a wife;
And as for nervous me, old pal,
Give me your own enchanting gal!'

Good CAPTAIN REECE, that worthy man,
Debated on his coxwain's plan:
'I quite agree,' he said, 'O BILL;
It is my duty, and I will.

'My daughter, that enchanting gurl,
Has just been promised to an earl,
And all my other familee,
To peers of various degree.

'But what are dukes and viscounts to
The happiness of all my crew?
The word I gave you I'll fulfil;
It is my duty, and I will.

'As you desire it shall befall,
I'll settle thousands on you all,
And I shall be, despite my hoard,
The only bachelor on board.'

The boatswain of *The Mantelpiece*,
He blushed and spoke to CAPTAIN REECE:
'I beg your honour's leave,' he said;
'If you would wish to go and wed,

'I have a widowed mother who
Would be the very thing for you—
She long has loved you from afar,
She washes for you, CAPTAIN R.'

The captain saw the dame that day—
Addressed her in his playful way—
'And did it want a wedding ring?
It was a tempting ickle sing!

'Well, well, the chaplain I will seek,
We'll all be married this day week—
At yonder church upon the hill;
It is my duty, and I will!'

The sisters, cousins, aunts, and niece,
And widowed ma of CAPTAIN REECE,
Attended there as they were bid;
It was their duty, and they did.

165 *A Thought from* Ruddigore

IF you wish in this world to advance,
Your merits you're bound to enhance,
You must stir it and stump it,
And blow your own trumpet,
Or, trust me, you haven't a chance!

166 *The Nightmare*

WHEN you're lying awake with a dismal headache, and repose is taboo'd by anxiety,
I conceive you may use any language you choose to indulge in, without impropriety;
For your brain is on fire—the bedclothes conspire of usual slumber to plunder you:
First your counterpane goes, and uncovers your toes, and your sheet slips demurely from under you;
Then the blanketing tickles—you feel like mixed pickles—so terribly sharp is the pricking,
And you're hot, and you're cross, and you tumble and toss till there's nothing 'twixt you and the ticking.
Then the bedclothes all creep to the ground in a heap, and you pick 'em all up in a tangle;
Next your pillow resigns and politely declines to remain at its usual angle!
Well, you get some repose in the form of a doze, with hot eye-balls and head ever aching,
But your slumbering teems with such horrible dreams that you'd very much better be waking;
For you dream you are crossing the Channel, and tossing about in a steamer from Harwich—

Which is something between a large bathing machine and a very small second-class carriage—
And you're giving a treat (penny ice and cold meat) to a party of friends and relations—
They're a ravenous horde—and they all came on board at Sloane Square and South Kensington Stations.
And bound on that journey you find your attorney (who started that morning from Devon);
He's a bit undersized, and you don't feel surprised when he tells you he's only eleven.
Well, you're driving like mad with this singular lad (by-the-bye the ship's now a four-wheeler),
And you're playing round games, and he calls you bad names when you tell him that 'ties pay the dealer';
But this you can't stand, so you throw up your hand, and you find you're as cold as an icicle,
In your shirt and your socks (the black silk with gold clocks), crossing Salisbury Plain on a bicycle:
And he and the crew are on bicycles too—which they've somehow or other invested in—
And he's telling the tars, all the particu*lars* of a company he's interested in—
It's a scheme of devices, to get at low prices, all goods from cough mixtures to cables
(Which tickled the sailors) by treating retailers, as though they were all vege*ta*bles—
You get a good spadesman to plant a small tradesman, (first take off his boots with a boot-tree),
And his legs will take root, and his fingers will shoot, and they'll blossom and bud like a fruit-tree—
From the greengrocer tree you get grapes and green pea, cauliflower, pineapple, and cranberries,
While the pastrycook plant, cherry brandy will grant, apple puffs, and three-corners, and banberries—
The shares are a penny, and ever so many are taken by Rothschild and Baring,
And just as a few are allotted to you, you awake with a shudder despairing—
You're a regular wreck, with a crick in your neck, and no wonder you snore, for your head's on the floor, and you've needles and pins from your soles to your shins, and your flesh is a-creep for your left leg's asleep, and you've cramp in your toes, and a fly on your nose, and some fluff in your lung, and a feverish tongue, and a thirst that's intense, and a general sense that you haven't been sleeping in clover;

But the darkness has passed, and it's daylight at last, and the night has been long—ditto ditto my song—and thank goodness they're both of them over!

BRET HARTE

1836–1902

167 *Plain Language from Truthful James*

WHICH I wish to remark—
And my language is plain—
That for ways that are dark,
And for tricks that are vain,
The heathen Chinee is peculiar,
Which the same I would rise to explain.

Ah Sin was his name;
And I will not deny
In regard to the same
What that name might imply;
But his smile it was pensive and childlike,
As I frequent remarked to Bill Nye.

It was August the third;
And quite soft was the skies:
Which it might be inferred
That Ah Sin was likewise;
Yet he played it that day upon William
And me in a way I despise.

Which we had a small game,
And Ah Sin took a hand.
It was Euchre. The same
He did not understand;
But he smiled as he sat by the table,
With a smile that was childlike and bland.

Yet the cards they were stocked
  In a way that I grieve,
And my feelings were shocked
  At the state of Nye's sleeve:
Which was stuffed full of aces and bowers,
  And the same with intent to deceive.

But the hands that were played
  By that heathen Chinee,
And the points that he made,
  Were quite frightful to see—
Till at last he put down a right bower,
  Which the same Nye had dealt unto me.

Then I looked up at Nye,
  And he gazed upon me;
And he rose with a sigh,
  And said, 'Can this be?
We are ruined by Chinese cheap labour—'
  And he went for that heathen Chinee.

In the scene that ensued
  I did not take a hand;
But the floor it was strewed
  Like the leaves on the strand
With the cards that Ah Sin had been hiding,
  In the game 'he did not understand.'

In his sleeves, which were long,
  He had twenty-four packs—
Which was coming it strong,
  Yet I state but the facts;
And we found on his nails, which were taper,
  What is frequent in tapers—that's wax.

Which is why I remark,
  And my language is plain,
That for ways that are dark,
  And for tricks that are vain,
The heathen Chinee is peculiar—
  Which the same I am free to maintain.

bowers] knaves, the highest cards in Euchre

THOMAS HARDY

1840–1928

168

Liddell and Scott
on the Completion of their Lexicon

'WELL, though it seems
Beyond our dreams,'
Said Liddell to Scott,
'We've really got
To the very end,
All inked and penned
Blotless and fair
Without turning a hair,
This sultry summer day, AD
Eighteen hundred and forty-three.

'I've often, I own,
Belched many a moan
At undertaking it,
And dreamt forsaking it.
—Yes, on to Pi,
When the end loomed nigh,
And friends said: "You've as good as done,"
I almost wished we'd not begun.
Even now, if people only knew
My sinkings, as we slowly drew
Along through Kappa, Lambda, Mu,
They'd be concerned at my misgiving,
And how I mused on a College living
Right down to Sigma,
But feared a stigma
If I succumbed, and left old Donnegan
For weary freshmen's eyes to con again:
And how I often, often wondered
What could have led me to have blundered
So far away from sound theology
To dialects and etymology;
Words, accents not to be breathed by men
Of any country ever again!'

'My heart most failed,
Indeed, quite quailed,'
Said Scott to Liddell,
'Long ere the middle! ...
'Twas one wet dawn
When, slippers on,
And a cold in the head anew,
Gazing at Delta
I turned and felt a
Wish for bed anew,
And to let supersedings
Of Passow's readings
In dialects go.
"That German has read
More than we!" I said;
Yea, several times did I feel so! ...

'O that first morning, smiling bland,
With sheets of foolscap, quills in hand,
To write ἀάατος and ἀαγής,
Followed by fifteen hundred pages,
What nerve was ours
So to back our powers,
Assured that we should reach ὠώδης
While there was breath left in our bodies!'

Liddell replied: 'Well, that's past now;
The job's done, thank God, anyhow.'
'And yet it's not,'
Considered Scott,
'For we've to get
Subscribers yet
We must remember;
Yes; by September.'

'O Lord; dismiss that. We'll succeed.
Dinner is my immediate need.
I feel as hollow as a fiddle,
Working so many hours,' said Liddell.

ANONYMOUS

169 *The Mid-West*

ACROSS those plains where once there roamed the Indian and the Scout,
The Swede with alcoholic breath sets rows of cabbage out.

MAX ADELER

1841–1915

170 *Willie*

WILLIE had a purple monkey climbing on a yellow stick,
And when he had sucked the paint all off it made him deadly sick;
And in his latest hours he clasped that monkey in his hand,
And bade good-bye to earth and went into a better land.

Oh no more he'll shoot his sister with his little wooden gun;
And no more he'll twist the pussy's tail and make her yowl for fun.
The pussy's tail now stands out straight; the gun is laid aside;
The monkey doesn't jump around since little Willie died.

EUGENE WARE

1841–1911

171 *Manila Bay*

O DEWEY was the morning upon the first of May.
And Dewey was the Admiral down in Manila Bay;
And Dewey were the Regent's eyes, them orbs of royal blue!
And Dewey feel discouraged? I Dew not think we Dew.

Manila Bay] Spanish–American War, 1898

GODFREY TURNER

fl. 1880

172

Synchoresis

WOULD you adopt a strong logical attitude,
  Bear this in mind, and, whatever you do,
Always allow your opponent full latitude,
  Whether or not his assumption be true.
Then, when he manifests feelings of gratitude
  Merely because you've not shut him up flat,
Turn his pet paradox into a platitude,
  With the remark, 'Why, *of course*, we know that!'
So, if you'd learn a good logical attitude,
  Keep this infallible maxim in view,
Always to grant your opponent full latitude,
  Whether or not his inductions be true.

Many an ass of a turn argumentative,
  Many a wiseacre, windy and dull,
Many a maniac tied to a tentative
  System that long ago turned out a mull.—
Many a bore, in short, loud though his patter is,
  Bent on the effort an issue to raise,
You may demolish, and silence his batteries,
  Just by agreement with all that he says;
That is to say, by adopting the attitude
  I've recommended so plainly to you,
Namely, to grant an opponent full latitude,
  Whether or not his assumptions are true.

Synchoresis] a rhetorical term meaning concession for the purpose of retort

GEORGE R. SIMS

1847–1922

173

A Garden Song

I SCORN the doubts and cares that hurt
  The world and all its mockeries,
My only care is now to squirt
  The ferns among my rockeries.

In early youth and later life
  I've seen an up and seen a down,
And now I have a loving wife
  To help me peg verbena down.

Of joys that come to womankind
  The loom of fate doth weave her few,
But here are summer joys entwined
  And bound with golden feverfew,

I've learnt the lessons one and all
  With which the world its sermon stocks,
Now, heedless of a rise or fall,
  I've Brompton and I've German stocks.

In peace and quiet pass our days,
  With nought to vex our craniums,
Our middle beds are all ablaze
  With red and white geraniums.

And like a boy I laugh when she,
  In Varden hat and Varden hose,
Comes slyly up the lawn at me
  To squirt me with the garden hose.

Let him who'd have the peace he needs
  Give all his worldly mumming up,
Then dig a garden, plant the seeds,
  And watch the product coming up.

ROBERT LOUIS STEVENSON

1850–1894

174

Epitaph

THE angler rose, he took his rod,
He kneeled and made his prayers to God.
The living God sat overhead:
The angler tripped, the eels were fed.

175

Good and Bad Children

CHILDREN, you are very little,
And your bones are very brittle;
If you would grow great and stately,
You must try to walk sedately.

You must still be bright and quiet,
And content with simple diet;
And remain, through all bewild'ring,
Innocent and honest children.

Happy hearts and happy faces,
Happy play in grassy places—
That was how, in ancient ages,
Children grew to kings and sages.

But the unkind and the unruly,
And the sort who eat unduly,
They must never hope for glory—
Theirs is quite a different story!

Cruel children, crying babies,
All grow up as geese and gabies,
Hated, as their age increases,
By their nephews and their nieces.

176

To Henry James

ADELA, Adela, Adela Chart
What have you done to my elderly heart?
Of all the ladies of paper and ink
I count you the paragon, call you the pink.

The word of your brother depicts you in part:
'You raving maniac!' Adela Chart;
But in all the asylums that cumber the ground,
So delightful a maniac was ne'er to be found.

I pore on you, dote on you, clasp you to heart,
I laud, love, and laugh at you, Adela Chart,
And thank my dear maker the while I admire
That I can be neither your husband nor sire.

Your husband's, your sire's were a difficult part;
You're a byway to suicide, Adela Chart;
But to read of, depicted by exquisite James,
O, sure you're the flower and quintessence of dames.

SAMUEL C. BUSHNELL

1852–1930

177

Boston

I COME from the city of Boston,
The home of the bean and the cod,
Where Cabots speak only to Lowells,
And Lowells speak only to God.

Adela Chart] heroine of Henry James's story *The Marriages*

ANONYMOUS

178 *The Ould Orange Flute*

IN the county Tyrone, in the town of Dungannon,
Where many a ruction myself had a han' in,
Bob Williamson lived, a weaver by trade
And all of us thought him a stout Orange blade.
On the twelfth of July as around it would come
Bob played on the flute to the sound of the drum,
You may talk of your harp, your piano or lute
But there's nothing compared to the ould Orange flute.

But Bob the deceiver he took us all in,
For he married a Papish called Brigid McGinn,
Turned Papish himself, and forsook the old cause
That gave us our freedom, religion and laws.
Now the boys of the place made some comment upon it,
And Bob had to fly to the Province of Connacht.
He fled with his wife and his fixings to boot,
And along with the latter his ould Orange flute.

At the chapel on Sundays, to atone for past deeds,
He said Paters and Aves and counted his beads,
Till after some time, at the priest's own desire,
He went with his ould flute to play in the choir.
He went with his ould flute to play for the Mass,
And the instrument shivered and said: 'Oh alas!'
And blow as he would, though it made a great noise,
The flute would play only 'The Protestant Boys'.

Bob jumped, and he started, and got in a flutter,
And threw his ould flute in the blest Holy Water;
He thought that this charm would bring some other sound
When he blew it again, it played 'Croppies lie down';
And for all he could whistle, and finger, and blow,
To play Papish music he found it no go;
'Kick the Pope', 'The Boyne Water', it freely would sound,
But one Papish squeak in it couldn't be found.

At the Council of priests that was held the next day,
They decided to banish the ould flute away

For they couldn't knock heresy out of its head
And they bought Bob a new one to play in its stead.
So the ould flute was doomed and its fate was pathetic,
'Twas fastened and burned at the stake as heretic,
While the flames roared around it they heard a strange noise
'Twas the ould flute still whistling 'The Protestant Boys'.

PERCY FRENCH

1854–1920

179 *The Queen's After-Dinner Speech*

(As Overheard and Cut into Lengths of Poetry by Jamesy Murphy, Deputy-Assistant-Waiter at the Viceregal Lodge)

'ME loving subjects,' sez she,
'Here's me best respects,' sez she,
'An' I'm proud this day,' sez she,
'Of the illigant way,' sez she,
'Ye gave me the hand,' sez she,
'Whin I came to land,' sez she.
'There was some people said,' sez she,
'They was greatly in dread,' sez she,
'I'd be murthered or shot,' sez she,
'As like as not,' sez she,
'But 'tis mighty clear,' sez she,
''Tis not over here,' sez she,
'I have cause to fear,' sez she.
''Tis them Belgiums,' sez she,
'That's throwin' bombs,' sez she,
'And scarin' the life,' sez she,
'Out o' me son and the wife,' sez she.
'But in these parts,' sez she,
'They have warrum hearts,' sez she,
'And they like me well,' sez she,
'Barrin' Anna Parnell,' sez she.
'I dunno, Earl,' sez she,
'What's come to the girl,' sez she,

The Queen's] Queen Victoria visited Ireland in 1900

'And that other wan,' sez she,
'That Maud Gonne,' sez she,
'Dhressin' in black,' sez she,
'To welcome me back,' sez she;
'Though I don't care,' sez she,
'What they wear,' sez she,
'An' all that gammon,' sez she,
'About me bringin' famine,' sez she.
'Now Maud 'ill write,' sez she,
'That I brought the blight,' sez she,
'Or altered the saysons,' sez she,
'For some private raysins,' sez she,
'An' I think there's a slate,' sez she,
'Off Willie Yeats,' sez she.
'He should be at home,' sez she,
'French polishin' a pome,' sez she,
'An' not writin' letters,' sez she,
'About his betters,' sez she,
'Paradin' me crimes,' sez she,
'In the "Irish Times",' sez she.
'But what does it matther,' sez she,
'This magpie chatther,' sez she,
'When that welcomin' roar,' sez she,
'Come up from the shore,' sez she,
'Right over the foam?' sez she,
''Twas like comin' home,' sez she,
'An' me heart fairly glowed,' sez she,
'Along the Rock Road,' sez she,
'An' by Merrion roun',' sez she,
'To Buttherstown,' sez she,
'Till I came to the ridge,' sez she
'Of the Leeson Street Bridge,' sez she,
'An' was welcomed in style,' sez she,
'By the beautiful smile,' sez she,
'Of me Lord Mayor Pile,' sez she.
'(Faith, if I done right,' sez she,
'I'd make him a knight,' sez she).
'Well, I needn't repeat,' sez she,
'How they cheered in each street,' sez she,
'Till I came to them lads,' sez she,
'Them "undergrads",' sez she.
'Indeed, an' indeed,' sez she,
'I've had many a God-speed,' sez she,
'But none to compare,' sez she,

'Wid what I got there,' sez she.
'Now pass the jug,' sez she,
'And fill up each mug,' sez she,
'Till I give ye a toast,' sez she,
'At which you may boast,' sez she.
'I've a power o' sons,' sez she,
'All sorts of ones,' says she:
'Some quiet as cows,' sez she,
'Some always in rows,' sez she,
'An' the one gives most trouble,' sez she,
'The mother loves double,' sez she,
'So drink to the min,' sez she,
'That have gone in to win,' sez she,
'And are clearin' the way,' sez she,
'To Pretoria to-day,' sez she.
'In the "Gap o' Danger",' sez she,
'There's a Connaught Ranger,' sez she,
'An' somewhere near,' sez she,
'Is a Fusilier,' sez she,
'An' the Inniskillings not far,' sez she,
'From the Heart o' the War,' sez she;
'An' I'll tell you what,' sez she,
'They may talk a lot,' sez she,
'And them Foreign Baboons,' sez she,
'May draw their cartoons,' sez she.
'But what they can't draw,' sez she,
'Is the Lion's claw,' sez she,
'And before our flag's furled,' sez she,
'We'll own the wurruld,' says she.

180 *The Mountains of Mourne*

OH, Mary, this London's a wonderful sight,
Wid the people here workin' by day and by night:
  They don't sow potatoes, nor barley, nor wheat,
  But there's gangs o' them diggin' for gold in the street—
At least, when I axed them, that's what I was told,
So I just took a hand at this diggin' for gold,
  But for all that I found there, I might as well be
  Where the Mountains o' Mourne sweep down to the sea.

I believe that, when writin', a wish you expressed
As to how the fine ladies in London were dressed.

Well, if you'll believe me, when axed to a ball,
They don't wear a top to their dresses at all!
Oh, I've seen them meself, and you could not, in thrath,
Say if they were bound for a ball or a bath—
Don't be startin' them fashions now, Mary Machree,
Where the Mountains o' Mourne sweep down to the sea.

I seen England's King from the top of a 'bus—
I never knew him, though he means to know us:
And though by the Saxon we once were oppressed,
Still, I cheered—God forgive me—I cheered wid the rest.
And now that he's visited Erin's green shore,
We'll be much better friends than we've been heretofore,
When we've got all we want, we're as quiet as can be
Where the Mountains o' Mourne sweep down to the sea.

You remember young Peter O'Loughlin, of course—
Well, here he is now at the head o' the Force.
I met him to-day, I was crossin' the Strand,
And he stopped the whole street wid wan wave of his hand:
And there we stood talking of days that are gone,
While the whole population of London looked on;
But for all these great powers, he's wishful like me,
To be back where dark Mourne sweeps down to the sea.

There's beautiful girls here—oh, never mind!
With beautiful shapes Nature never designed,
And lovely complexions, all roses and crame,
But O'Loughlin remarked wid regard to them same:
'That if at those roses you venture to sip,
The colour might all come away on your lip,'
So I'll wait for the wild rose that's waitin' for me—
Where the Mountains o' Mourne sweep down to the sea.

J. K. STEPHEN

1859–1892

181 *On a Rhine Steamer*

REPUBLIC of the West,
  Enlightened, free, sublime,
Unquestionably best
  Production of our time.

The telephone is thine,
  And thine the Pullman Car,
The caucus, the divine
  Intense electric star.

To thee we likewise owe
  The venerable names
Of Edgar Allan Poe,
  And Mr Henry James.

In short it's due to thee,
  Thou kind of Western star,
That we have come to be
  Precisely what we are.

But every now and then,
  It cannot be denied,
You breed a kind of men
  Who are not dignified,

Or courteous or refined,
  Benevolent or wise,
Or gifted with a mind
  Beyond the common size,

Or notable for tact,
  Agreeable to me,
Or anything, in fact,
That people ought to be.

182

Malines

(*Midnight, July 4th, 1882*)

BELGIAN, with cumbrous tread and iron boots
Who in the murky middle of the night
Designing to renew the foul pursuits
In which thy life is passed, ill-favoured wight,
And wishing on the platform to alight
Where thou couldst mingle with thy fellow-brutes
Didst walk the carriage floor (a leprous sight),
As o'er the sky some baleful meteor shoots:
Upon my slippered foot thou didst descend,
Didst rouse me from my slumbers mad with pain,
And laughedst aloud for several minutes' space.
Oh may'st thou suffer tortures without end:
May fiends with glowing pincers rend thy brain,
And beetles batten on thy blackened face!

SIR ARTHUR CONAN DOYLE

1859–1930

183

To An Undiscerning Critic

SURE, there are times when one cries with acidity,
'Where are the limits of human stupidity?'
Here is a critic who says as a platitude,
That I am guilty because 'in ingratitude,
Sherlock, the sleuth hound, with motives ulterior,
Sneers at Poe's Dupin as very "inferior".'

Have you not learned, my esteemed commentator,
That the created is not the creator?
As the creator I've praised to satiety
Poe's Monsieur Dupin, his skill and variety,
And have admitted that in my detective work,
I owe to my model a deal of selective work.

But is it not on the verge of inanity
To put down to me my creation's crude vanity?
He, the created, the puppet of fiction,
Would not brook rivals nor stand contradiction,
He, the created, would scoff and would sneer,
Where I, the Creator, would bow and revere.

So please grip this fact with your cerebral tentacle,
The doll and its maker are never identical.

J. W. MACKAIL (1859–1945) and CECIL SPRING-RICE (1859–1918)

184

On the Hon. George Nathaniel Curzon, Commoner of Balliol

MY name is George Nathaniel Curzon,
I am a most superior person.
My cheeks are pink, my hair is sleek,
I dine at Blenheim twice a week.

A. E. HOUSMAN

1859–1936

185

The Shades of Night

THE shades of night were falling fast,
And the rain was falling faster;
When through an Alpine village passed
An Alpine village pastor:
A youth who bore mid snow and ice
A bird that wouldn't chirrup,
And a banner with the strange device—
'Mrs Winslow's soothing syrup.'

'Beware the pass,' the old man said,
  'My bold, my desperate fellah;
Dark lowers the tempest overhead,
  And you'll want your umberella;
And the roaring torrent is deep and wide—
  You may hear how loud it washes.'
But still that clarion voice replied:
  'I've got my old goloshes.'

'Oh, stay,' the maiden said, 'and rest
  (For the wind blows from the nor'ward)
Thy weary head upon my breast—
  And please don't think I'm forward.'
A tear stood in his bright blue eye,
  And he gladly would have tarried;
But still he answered with a sigh:
  'Unhappily I'm married.'

186

Fragment of an English Opera

(*Designed as a model for young librettists*)

Dramatis personae:
Father (bass)
Mother (contralto)
Daughter (soprano)

Scene: A room. *Time*: Evening

Father RETIRE, my daughter;
  Prayers have been said;
Take your warm water
  And go to bed.
Daughter But I had rather
  Sit up instead.
Father I am your father,
  So go to bed.
Daughter Are you my father?
Father I think so, rather:
  You go to bed.
Mother My daughter, vanish;
  You hear me speak:
This is not Spanish,
  Nor is it Greek.

Daughter	Oh, what a bother!
	Would I were dead!
Mother	I am your mother,
	So go to bed.
Daughter	Are you my mother?
Mother	You have no other:
	You go to bed.
Father	Take your bed-candle
	And take it quick.
	This is the handle.
Daughter	Is *this* the handle?
Father	No, that's the wick.
	This is the handle,
	At this end here.
	Take your bed-candle
	And disappear.
Daughter	Oh dear, oh dear!
Father & Mother	Take your warm water,
	As we have said;
	You are our daughter,
	So go to bed.
Daughter	Am I your daughter?
Father & Mother	If not, you oughter:
	You go to bed.
Daughter	I am their daughter;
	If not, I oughter:
	Prayers have been said.
	This is my mother;
	I have no other:
	Would I were dead!
	That is my father;
	He thinks so, rather:
	Oh dear, oh dear!
	I take my candle;
	This is the handle:
	I disappear.
Father & Mother	The coast is clear.

187

The Pope

It is a fearful thing to be
    The Pope.
That cross will not be laid on me,
    I hope.
A righteous God would not permit
    It.
The Pope himself must often say,
After the labours of the day,
'It is a fearful thing to be
    Me.'

EDGAR BATEMAN

1860–1946

188

It's a Great Big Shame

I've lost a pal, 'e's the best in all the tahn,
But don't you fink 'im dead, becos 'e aint—
But since 'e's wed 'e 'as 'ad ter knuckle dahn—
It's enuf to wex the temper of a saint!
'E's a brewer's drayman wiv a leg o' mutton fist,
An' as strong as a bullick or an 'orse—
Yet in 'er 'ands 'e's like a little kid—
Oh! I wish as I could get 'im a divorce.

Chorus:
It's a great big shame, an' if she belonged ter me
I'd let 'er know who's who—
Naggin' at a feller wot is six foot free,
And 'er only four foot two!
Oh! they 'adn't been married not a month nor more,
When underneath her fumb goes Jim—
Oh, isn't it a pity as the likes of 'er
Should put upon the likes of 'im?

Now Jim was class 'e could sing a decent song,
And at scrappin' 'e 'ad won some great renown;
It took two coppers for ter make 'im move along,
And annuvver six to 'old the feller dahn.
But today when I axes would 'e come an' 'ave some beer,
To the doorstep on tip-toe 'e arrives;
'I daresn't,' says 'e 'Don't shout, cos she'll 'ear
I've got ter clean the windows an' the knives.'

On a Sunday morn, wiv a dozen pals or more,
'E'd play at pitch and toss along the Lea;
But now she bullies 'im a scrubbin' 'o the floor
Such a change, well, I never did see.
Wiv a apron on 'im, I twigged 'im, on 'is knees
A rubbin' up the old 'arf stone;
Wot wiv emptyin' the ashes and a-shellin' o' the peas,
I'm blowed if 'e can call 'is self 'is own!

It's a great big shame, &c.

SIR WALTER RALEIGH

1861–1922

189 *Wishes of an Elderly Man, Wished at a Garden Party, June 1914*

I WISH I loved the Human Race;
I wish I loved its silly face;
I wish I loved the way it walks;
I wish I loved the way it talks;
And when I'm introduced to one
I wish I thought *What Jolly Fun*!

190 *Lines Suggested by an Edition of Blake's Poems*

IF you try to do what's right
You pass your life in a horrible fright,
And your Emanation—Lord protect her!—
Commits adultery with your Spectre.

SIR ARTHUR QUILLER-COUCH

1863–1944

191

The Harbour of Fowey

O THE Harbour of Fowey
    Is a beautiful spot,
And it's there I enjowey
    To sail in a yot;
Or to race in a yacht
    Round a mark or a buoy—
Such a beautiful spacht
    Is the Harbour of Fuoy!

But the wave mountain-high,
    And the violent storm,
Do I risk them? Not Igh!
    But prefer to sit worm
With a book on my knees
    By the library fire,
While I list to the brees
    Rising hire and hire.

And my leisure 's addressed
    To composing of verse
Which, if hardly the bessed,
    Might be easily werse.
And, the spelling I use
    Should the critics condemn,
Why, I have my own vuse
    And I don't think of themn.

Yes, I have my own views:
  But the teachers I follow
Are the Lyrical Miews
  And the Delphic Apollow.
Unto them I am debtor
  For spelling and rhyme,
And I'm doing it bebtor
  And bebtor each thyme.

ERNEST LAWRENCE THAYER

1863–1940

192 *Casey at the Bat*

THE outlook wasn't brilliant for the Mudville nine that day;
The score stood four to two with but one inning more to play.
So when Cooney died at second, and Burrows did the same,
A pallor wreathed the features of the patrons of the game.
A straggling few got up to go in deep despair. The rest
Clung to the hope which springs eternal in the human breast;
They thought, 'If only Casey could but get a whack at that—
We'd put up even money now with Casey at the bat.'
But Flynn preceded Casey, as did also Jimmy Blake,
And the former was a lulu and the latter was a fake;
So upon that stricken multitude a deathlike silence sat,
For there seemed but little chance of Casey's getting to the bat.
But Flynn let drive a single, to the wonderment of all,
And Blake, the much despis-ed, tore the cover off the ball;
And when the dust had lifted, and the men saw what had occurred,
There was Jimmy safe at second and Flynn a-hugging third.
Then from five thousand throats and more there rose a lusty yell;
It rumbled in the mountaintops, it rattled in the dell;
It knocked upon the hillside and recoiled upon the flat,
For Casey, mighty Casey, was advancing to the bat.
There was ease in Casey's manner as he stepped into his place;
There was pride in Casey's bearing and a smile on Casey's face.
And when, responding to the cheers, he lightly doffed his hat,
No stranger in the crowd could doubt 'twas Casey at the bat.
Ten thousand eyes were on him as he rubbed his hands with dirt;
Five thousand tongues applauded when he wiped them on his shirt.
Then while the writhing pitcher ground the ball into his hip,
Defiance gleamed in Casey's eye, a sneer curled Casey's lip.
And now the leather-covered sphere came hurtling through the air,
And Casey stood a-watching it in haughty grandeur there.
Close by the sturdy batsman the ball unheeded sped—
'That ain't my style,' said Casey—'Strike one,' the Umpire said.
From the benches black with people, there went up a muffled roar,
Like the beating of the storm-waves on a stern and distant shore.
'Kill him! kill the umpire!' shouted someone on the stand;
And it's likely they'd have killed him had not Casey raised his hand.

With a smile of Christian charity great Casey's visage shone;
He stilled the rising tumult; he bade the game go on;
He signalled to the pitcher, and once more the spheroid flew;
But Casey still ignored it, and the Umpire said, 'Strike two.'
'Fraud!' cried the maddened thousands, and the echo answered, 'Fraud!'
But one scornful look from Casey and the multitude was awed.
They saw his face grow stern and cold, they saw his muscles strain,
And they knew that Casey wouldn't let that ball go by again.
The sneer is gone from Casey's lip, his teeth are clenched in hate;
He pounds with cruel violence his bat upon the plate.
And now the pitcher holds the ball, and now he lets it go,
And now the air is shattered by the force of Casey's blow.
Oh, somewhere in this favored land the sun is shining bright;
The band is playing somewhere, and somewhere hearts are light,
And somewhere men are laughing, and somewhere children shout;
But there is no joy in Mudville—mighty Casey has struck out.

OLIVER HERFORD

1863–1939

193

The Smile of the Walrus

THE Smile of the Walrus is wild and distraught,
And tinged with pale purples and greens,
Like the Smile of a Thinker who thinks a Great Thought
And isn't quite sure what it means.

194

The Smile of the Goat

THE Smile of the Goat has a meaning that few
Will mistake, and explains in a measure
The Censor attending a risqué Revue
And combining Stern Duty with pleasure.

RUDYARD KIPLING

1865–1936

195

The Sergeant's Weddin'

'E WAS warned agin 'er—
  That's what made 'im look:
She was warned agin' 'im—
  That is why she took.
Wouldn't 'ear no reason,
  Went an' done it blind;
We know all about 'em,
  They've got all to find!

Cheer for the Sergeant's weddin'—
  Give 'em one cheer more!
Grey gun-'orses in the lando,
  An' a rogue is married to, etc.

What's the use o' tellin'
  'Arf the lot she's been?
'E's a bloomin' robber,
  An' 'e keeps canteen.
'Ow did 'e get 'is buggy?
  Gawd, you needn't ask!
Made 'is forty gallon
  Out of every cask!

Watch 'im, with 'is 'air cut,
  Count us filin' by—
Won't the Colonel praise 'is
  Pop—u—lar—i—ty!
We 'ave scores to settle—
  Scores for more than beer;
She's the girl to pay 'em—
  That is why we're 'ere!

See the Chaplain thinkin'?
  See the women smile?
Twig the married winkin'
  As they take the aisle?

Keep your side-arms quiet,
   Dressin' by the Band.
Ho! You 'oly beggars,
   Cough be'ind your 'and!

Now it's done an' over,
   'Ear the organ squeak,
'*Voice that breathed o'er Eden*'—
   Ain't she got the cheek!
White an' laylock ribbons,
   Think yourself so fine!
I'd pray Gawd to take yer
   'Fore I made yer mine!

Escort to the kerridge,
   Wish 'im luck, the brute!
Chuck the slippers after—
   (Pity 'tain't a boot!)
Bowin' like a lady,
   Blushin' like a lad—
'Oo would say to see 'em
   Both is rotten bad?

Cheer for the Sergeant's weddin'—
   Give 'em one cheer more!
Grey gun-'orses in the lando,
   An' a rogue is married to, etc.

196 *Road-Song of the* Bandar-Log

HERE we go in a flung festoon,
Half-way up to the jealous moon!
Don't you envy our pranceful bands?
Don't you wish you had extra hands?
Wouldn't you like if your tails were—*so*—
Curved in the shape of a Cupid's bow?
      Now you're angry, but—never mind,
      Brother, thy tail hangs down behind!

Here we sit in a branchy row,
Thinking of beautiful things we know;

Bandar-Log] the Monkey-People

Dreaming of deeds that we mean to do,
All complete, in a minute or two—
Something noble and grand and good,
Won by merely wishing we could.
    Now we're going to—never mind,
    Brother, thy tail hangs down behind!

All the talk we ever have heard
Uttered by bat or beast or bird—
Hide or fin or scale or feather—
Jabber it quickly and all together!
Excellent! Wonderful! Once again!
Now we are talking just like men.
    Let's pretend we are ... Never mind!
    Brother, thy tail hangs down behind!
    This is the way of the Monkey-kind!

ANONYMOUS

197 *Two Headmistresses*

MISS Buss and Miss Beale
Cupid's darts do not feel.
How different from us
Are Miss Beale and Miss Buss.

ANONYMOUS

198 *I Was Born Almost Ten Thousand Years Ago*

I WAS born almost ten thousand years ago,
And there's nothing in the world that I don't know;
I saw Peter, Paul and Moses,
Playing ring-around-the-roses
And I'm here to lick the guy what says 'taint so.

I saw Samson when he laid the village cold,
Saw Daniel tame the lions in the hold,
And helped build the Tower of Babel,
Up as high as they were able,
And there's lots of other things I haven't told.

I taught Solomon his little A-B-C's,
I helped Brigham Young to make Limburger cheese,
And while sailing down the bay
With Methusaleh one day,
I saved his flowing whiskers from the breeze.

Queen Elizabeth she fell in love with me
We were married in Milwaukee secretly,
But I schemed around and shook her,
And I went with General Hooker
To shoot mosquitoes down in Tennessee.

I remember when the country had a king,
I saw Cleopatra pawn her wedding ring,
And I saw the flags a-flying
When George Washington stopped lying,
On the night when Patti first began to sing.

ANONYMOUS

199

Lydia Pinkham

THEN we'll sing of Lydia Pinkham,
And her love for the human race;
How she sold her veg'table compound
And the papers publish'd her face.

Oh, it sells for a dollar a bottle
Which is very cheap you see,
And if it doesn't cure you
She will sell you six for three.

GELETT BURGESS

1866–1951

200

The Purple Cow

I NEVER saw a Purple Cow,
I never hope to see one;
But I can tell you, anyhow,
I'd rather see than be one.

201

Trapping Fairies

TRAPPING fairies in West Virginia:
I think I never saw fairies skinnier.

202

Leave-Taking

THE proper way to leave a room
Is not to plunge it into gloom;
Just make a Joke before you go,
And then escape before they know.

203

Cinq Ans Après

AH, yes! I wrote the 'Purple Cow'—
I'm Sorry, now, I Wrote it!
But I can Tell you, Anyhow,
I'll Kill you if you Quote it!

GEORGE ADE

1866–1944

204

R-E-M-O-R-S-E

THE cocktail is a pleasant drink,
It's mild and harmless, I don't think.
When you've had one, you call for two,
And then you don't care what you do.
Last night I hoisted twenty-three
Of these arrangements into me;
My wealth increased, I swelled with pride;
I was pickled, primed and ossified.

R-E-M-O-R-S-E!
Those dry martinis did the work for me;
Last night at twelve I felt immense;
Today I feel like thirty cents.
At four I sought my whirling bed,
At eight I woke with such a head!
It is no time for mirth or laughter—
The cold, gray dawn of the morning after.

If ever I want to sign the pledge,
It's the morning after I've had an edge;
When I've been full of the oil of joy
And fancied I was a sporty boy.
This world was one kaleidoscope
Of purple bliss, transcendent hope.
But now I'm feeling mighty blue—
Three cheers for the WCTU!

R-E-M-O-R-S-E!
The water wagon is the place for me;
I think that somewhere in the game,
I wept and told my maiden name.
My eyes are bleared, my coppers hot;
I try to eat, but I can not;
It is no time for mirth or laughter—
The cold, gray dawn of the morning after.

WCTU] Women's Christian Temperance Union

E. G. MURPHY ('DRYBLOWER')

1867–1939

205 *'Thank you, Mr Rason, for the Apples'*

London item: 'At the Buckingham Palace banquet, given to the Colonial Premiers and others, Westralian apples were eaten by the King. He afterwards conversed with Westralia's Agent-General (C. H. Rason), and on parting said, "Thank you, Mr Rason, for the apples" '.

THE banquet was a bonza, a rare recherché feed,
  Every item there was up-to-date and swagger,
From the appetizing oysters to the coffee and the weed,
  When the guests had waded through the Royal jagger.
There were squatters from the Sydneyside and butter kings from Vic.,
  Lack of space, alone, from naming them prevents us;
But the proudest pea of all who lapped the pale Imperial shick
  Was Rason, who in London represents us.
The King was kind and courteous, and had grins for every guest,
  With that tactfulness that always grips and grapples,
But to Hector's lug delighted came this message bright and blest—
  'Thank you, Mr Rason, for the apples!'

The assemblage was a brilliant one, for Deakin, silver-tongued,
  Had come to loose his oratory thrilling,
They hiccoughed Gord-Save-Edward with a chorus leather-lunged,
  What time the loyal liquor they were swilling.
They speeched of silver kinships, and of hands across the sea,
  They preached the same old platitudes redundant.
For whenever there's a plenitude of amber eau-de-vie,
  Cronk sentiment is equally abundant;
They boasted how the Empire's Sons would rally round the Flag,
  When Britain's drum reverberates and rapples,
But the ear of Cornthwaite Hector had no room for any gag
  Than 'Thank you, Mr Rason, for the apples!'

They were grown in boggy Boyanup, and washed by Darling dews,
  They were picked by cockies' kiddies fat and freckled,
Who little thought they'd ever mix with Nedward's banquet booze,
  When the Royal chef a score of flunkies heckled.

They were shipped aboard a liner with the butter and the beef,
For weeks they heard the engine's diapason,
Till they reached where London's swindle-shark and market-rigging thief
Rubs ribs with West Australia's only Rason.
They graced the Royal blow-out where the loyal speakers wreathed,
His pappycock and persiflage and papples,
Till in Cornthwaite Hector's listener Ned the Seventh softly breathed,
'Thank you, Mr Rason, for the apples.'

Little thought old sleepy Boyanup, of whom the world scarce heard,
Where the poddy-calf pursues its mooing mummy,
That the product of its orchard would in London be interred
In a real, live, reigning, Royal Monarch's tummy,
Little thought the cockies' kiddies, who each morn the moo-cow drain
Of milk that nearer blue than creamy-white is,
That the apples they'd exported would in time allay the pain
In a British Sovereign's paunch appendicitis,
Little thought they that an apple-core from Boyanup serene,
Might put Royalty to sleep in storied chapels,
Little thought our Agent-General would hear with blushing mien—
'Thank you, Mr Rason, for the apples!'

But supposing, just supposing, that the fruit from Boyanup
Upset the weak digestion of King Neddy,
Imagine how the household would be telephoning up
To get the surgeon's apparatus ready,
Imagine Alexandra mixing mustard in a pan
To put upon the Monarch's Little Mary,
Imagine how her feet would feel as round about she ran,
In a nightie rather delicate and airy;
It isn't very cosy to be wakened from a dream,
When the cold, grey dawn the Eastern darkness dapples,
While a King in fruit-convulsions mutters through the poultice-steam—
'Blast you, Mr Rason, and your apples!'

ANONYMOUS

206 *In the Days of Old Rameses*

IN the days of old Rameses, are you on, are you on?
They told the same thing, the very same thing.
In the days of old Rameses, that story had paresis,
Are you on, are you on, are you on?

Adam told it to the beast before the fall, are you on?
He told the same thing, the very same thing.
When he told it to the creatures, it possessed redeeming features,
But to tell it now requires a lot of gall.

In the days of Sodom and Gomorrah, are you on?
They told the same thing, the very same thing;
In Sodom and Gomorrah, people told it to their sorrow,
Are you on, are you on, are you on?

In the days of ancient Florence, are you on?
They told the same thing, the very same thing;
In the days of ancient Florence, it was held in great abhorrence,
Are you on, are you on, are you on?

CHARLES INGE

1868–1957

207 *On Professor Coué*

THIS very remarkable man
Commends a most practical plan:
You can do what you want
If you don't think you can't,
So don't think you can't think you can.

HILAIRE BELLOC

1870–1953

208 *Henry King, Who Chewed Bits of String, and Was Early Cut Off in Dreadful Agonies*

THE Chief Defect of Henry King
Was chewing little bits of String.
At last he swallowed some which tied
Itself in ugly Knots inside.
Physicians of the Utmost Fame
Were called at once; but when they came
They answered, as they took their Fees,
'There is no Cure for this Disease.
Henry will very soon be dead.'
His Parents stood about his Bed
Lamenting his Untimely Death,
When Henry, with his Latest Breath,
Cried—'Oh, my Friends, be warned by me,
That Breakfast, Dinner, Lunch, and Tea
Are all the Human Frame requires ...'
With that, the Wretched Child expires.

209 *The Pacifist*

PALE Ebenezer thought it wrong to fight,
But Roaring Bill (who killed him) thought it right.

210 *On Mundane Acquaintances*

GOOD morning, Algernon: Good morning, Percy.
Good morning, Mrs Roebeck. Christ have mercy!

211 *Obiter Dicta*

I

SIR HENRY WAFFLE KC (*continuing*)

SIR ANTHONY HABBERTON, Justice and Knight,
Was enfeoffed of two acres of land
And it doesn't sound much till you hear that the site
Was a strip to the South of the Strand.

HIS LORDSHIP (*Obiter Dictum*)

A strip to the South of the Strand
Is a good situation for land.
It is healthy and dry
And sufficiently high
And convenient on every hand.

II

SIR HENRY WAFFLE KC (*continuing*)

Now Sir Anthony, shooting in Timberley Wood,
Was imprudent enough to take cold;
And he died without warning at six in the morning,
Because he was awfully old.

HIS LORDSHIP (*Obiter Dictum*)

I have often been credibly told
That when people are awfully old
Though cigars are a curse
And strong waters are worse
There is nothing so fatal as cold.

III

SIR HENRY WAFFLE KC (*continuing*)

But Archibald answered on hearing the news:
'I never move out till I must'.
Which was all very jolly for *Cestui que Use*
But the Devil for *Cestui que Trust*.

HIS LORDSHIP (*Obiter Dictum*)

The office of *Cestui que Trust*
Is reserved for the learned and just.
Any villain you choose
May be *Cestui que Use*,
But a Lawyer for *Cestui que Trust*.

IV

SIR HENRY WAFFLE KC (*continuing*)

Now the ruling laid down in *Regina v. Brown*
May be cited ...

HIS LORDSHIP (*rising energetically*)

You're wrong! It may not!
I've strained all my powers
For some thirty-six hours
To unravel this pestilent rot.

THE WHOLE COURT (*rising and singing in chorus*)

Your Lordship is sound to the core.
It is nearly a quarter to four.
We've had quite enough of this horrible stuff
And we don't want to hear any more!

LITTLE SILLY MAN (*rising at the back of the Court*)

Your Lordship is perfectly right.
He can't go on rhyming all night.
I suggest ...
(*He is gagged, bound and dragged off to a Dungeon*).

212

Is There Any Reward?

Is there any reward?
  I'm beginning to doubt it.
I am broken and bored,
  Is there any reward?
Reassure me, Good Lord,
  And inform me about it.
Is there any reward?
  I'm beginning to doubt it.

213

On Lady Poltagrue: A Public Peril

The Devil, having nothing else to do,
Went off to tempt My Lady Poltagrue.
My Lady, tempted by a private whim,
To his extreme annoyance, tempted him.

214

Imitation

Hen THEREFORE do thou, stiff-set Northumberland,
Retire to Chester, and my cousin here,
The noble Bedford, hie to Glo'ster straight
And give our Royal ordinance and word
That in this fit and strife of empery
No loss shall stand account. To this compulsion
I pledge my sword, my person and my honour
On the Great Seal of England: so farewell.
Swift to your charges: nought was ever done
Unless at some time it were first begun.

215

Lord Heygate

LORD HEYGATE had a troubled face,
His furniture was commonplace—
The sort of Peer who well might pass
For someone of the middle class.
I do not think you want to hear
About this unimportant Peer.

216

The World's a Stage

THE world's a stage. The trifling entrance fee
Is paid (by proxy) to the registrar.
The Orchestra is very loud and free
But plays no music in particular.
They do not print a programme, that I know.
The cast is large. There isn't any plot.
The acting of the piece is far below
The very worst of modernistic rot.

The only part about it I enjoy
Is what was called in English the Foyay.
There will I stand apart awhile and toy
With thought, and set my cigarette alight;
And then—without returning to the play—
On with my coat and out into the night.

A. H. SIDGWICK

fl. 1900

217 *The Strenuous Life*

ON the cabin-roof I lie
Gazing into vacancy.
Make no noise and break no jest,
I am peaceful and at rest.

Somewhere back in days gone by
I did something—was it I?
Do not ask: I have forgot
Whether it was I or not.

Sometime I may have to do
Something else; but so may you.
Do not argue, but admit
That we need not think of it.

Thought has ever been my foe;
That is so. Yes. That is so.
On the cabin-roof I lie
Gazing into vacancy.

J. M. SYNGE

1871–1909

218 *The Curse*

To a sister of an enemy of the author's who disapproved of 'The Playboy'

LORD, confound this surly sister,
Blight her brow with blotch and blister,
Cramp her larynx, lung, and liver,
In her guts a galling give her.

Let her live to earn her dinners
In Mountjoy with seedy sinners:
Lord, this judgment quickly bring,
And I'm your servant, J. M. Synge.

ARTHUR GUITERMAN

1871–1943

219 *Everything in its Place*

THE skeleton is hiding in the closet as it should,
The needle's in the haystack and the trees are in the wood,
The fly is in the ointment and the froth is on the beer,
The bee is in the bonnet and the flea is in the ear.

The meat is in the coconut, the cat is in the bag,
The dog is in the manger and the goat is on the crag,
The worm is in the apple and the clam is on the shore,
The birds are in the bushes and the wolf is at the door.

SIR MAX BEERBOHM

1872–1956

220 *On the Imprint of the First English Edition of* The Works of Max Beerbohm

'London: JOHN LANE, *The Bodley Head*
New York: CHARLES SCRIBNER'S SONS.'
This plain announcement, nicely read,
Iambically runs.

221 *Chorus of a Song that Might Have Been Written by Albert Chevalier*

I drops in to see young Ben
In 'is tap-room now an' then,
And I likes to see 'im gettin' on becoz
'E's got pluck and 'e's got brains,
And 'e takes no end o' pains,
But—'e'll never be the man 'is Father woz.

222 *After Hilaire Belloc*

At dawn to-morrow
On Storrington Barrow
I'll beg or borrow
A bow and arrow
And shoot sleek sorrow
Through the marrow.
The floods are out and the ford is narrow,
The stars hang dead and my limbs are lead,
But ale is gold
And there's good foot-hold
On the Cuckfield side of Storrington Barrow.

223 *In a Copy of More's (or Shaw's or Wells's or Plato's or Anybody's)* Utopia

So this is Utopia, is it? Well
I beg your pardon, I thought it was Hell.

224 *Addition to Kipling's 'The Dead King (Edward VII), 1910'*

Wisely and well was it said of him, 'Hang it all, he's a
Mixture of Jesus, Apollo, Goliath and Julius Caesar!'
Always he plans as an ever Do-Right-man, never an Err-man,
And never a drop of the blood in his beautiful body was German.
'God save him,' we said when he lived, but the words now sound odd,
For we know that in Heaven above at this moment *he*'s saving *God*.

225 from *Old Surrey Saws and Sayings*

Collected and communicated by Sir Max Beerbohm, PRA (Professor of Rural Archaeology)

A RED sky at night
Is a shepherd's delight,
A red sky in the morning
Is a shepherd's warning,
A sky that looks bad
Is a shepherd's plaid,
A good-looking sky
Is a shepherd's pie.

WALTER DE LA MARE

1873–1956

226 *Moonshine*

THERE was a young lady of Rheims,
There was an old poet of Gizeh;
He rhymed on the deepest and sweetest of themes,
She scorned all his efforts to please her:
And he sighed, 'Ah, I see,
She and sense won't agree.'
So he scribbled her moonshine, mere moonshine, and she,
With jubilant screams, packed her trunk up in Rheims,
Cried aloud, 'I am coming, O Bard of my dreams!'
And was clasped to his bosom in Gizeh.

227 *Dear Sir*

THERE was an old Rabbi of Ur;
He loved a Miss Beaulieu.
She sent him a letter: 'Dear Sir ...'
Then a stone-cold 'Yours truly.'

Now what she could mean
By the dots in between
Is not plain to be seen.
We can but infer the Rabbi of Ur
Enquired of Miss Beaulieu.

228

The Shubble

THERE was an old man said, 'I fear
That life, my dear friends, is a bubble,
Still, with all due respect to a Philistine ear,
A limerick's best when it's double.'
When they said, 'But the waste
Of time, temper, taste!'
He gulped down his ink with cantankerous haste,
And chopped off his head with a shubble.

229

Pooh!

DAINTY Miss Apathy
Sat on a sofa,
Dangling her legs,
And with nothing to do;
She looked at a drawing of
Old Queen Victoria,
At a rug from far Persia—
An exquisite blue;
At a bowl of bright tulips;
A needlework picture
Of doves caged in wicker
You could almost hear coo;
She looked at the switch
That evokes e-
Lectricity;
At the coals of an age
BC millions and two—
When the trees were like ferns
And the reptiles all flew;
She looked at the cat
Asleep on the hearthrug,
At the sky at the window,—
The clouds in it, too;

And a marvellous light
From the West burning through:
And the one silly word
In her desolate noddle
As she dangled her legs,
Having nothing to do,
Was not, as you'd guess,
Of dumfoundered felicity,
But contained just four letters,
And these pronounced *POOH*!

ANONYMOUS

230

Elinor Glyn

WOULD you like to sin
With Elinor Glyn
On a tiger-skin?
Or would you prefer
to err
with her
on some other fur?

G. K. CHESTERTON

1874–1936

231

A Ballade of Suicide

THE gallows in my garden, people say,
Is new and neat and adequately tall.
I tie the noose on in a knowing way
As one that knots his necktie for a ball;
But just as all the neighbours—on the wall—
Are drawing a long breath to shout 'Hurray!'
The strangest whim has seized me ... After all
I think I will not hang myself to-day.

To-morrow is the time I get my pay—
My uncle's sword is hanging in the hall—
I see a little cloud all pink and grey—
Perhaps the Rector's mother will *not* call—
I fancy that I heard from Mr Gall
That mushrooms could be cooked another way—
I never read the works of Juvenal—
I think I will not hang myself to-day.

The world will have another washing day;
The decadents decay; the pedants pall;
And H. G. Wells has found that children play,
And Bernard Shaw discovered that they squall;
Rationalists are growing rational—
And through thick woods one finds a stream astray,
So secret that the very sky seems small—
I think I will not hang myself to-day.

ENVOI

Prince, I can hear the trumpet of Germinal,
The tumbrils toiling up the terrible way;
Even to-day your royal head may fall—
I think I will not hang myself to-day.

232 *The Fat White Woman Speaks*

O fat white woman whom nobody loves,
Why do you walk through the fields in gloves,
Missing so much and so much?

FRANCES CORNFORD

WHY do you rush through the field in trains,
Guessing so much and so much.
Why do you flash through the flowery meads,
Fat-head poet that nobody reads;
And why do you know such a frightful lot
About people in gloves as such?
And how the devil can you be sure,
Guessing so much and so much,
How do you know but what someone who loves
Always to see me in nice white gloves
At the end of the field you are rushing by,
Is waiting for his Old Dutch?

233 *On Reading 'God'*

(Mr Middleton Murry explains that his book with this title records his farewell to God)

MURRY, on finding *le Bon Dieu*
Chose difficile à croire
Illogically said 'Adieu,'
But God said 'Au Revoir.'

234 *The Aristocrat*

THE Devil is a gentleman, and asks you down to stay
At his little place at What'sitsname (it isn't far away).
They say the sport is splendid; there is always something new,
And fairy scenes, and fearful feats that none but he can do;
He can shoot the feathered cherubs if they fly on the estate,
Or fish for Father Neptune with the mermaids for a bait;
He scaled amid the staggering stars that precipice, the sky,
And blew his trumpet above heaven, and got by mastery
The starry crown of God Himself, and shoved it on the shelf;
But the Devil is a gentleman, and doesn't brag himself.

O blind your eyes and break your heart and hack your hand away,
And lose your love and shave your head; but do not go to stay
At the little place in What'sitsname where folks are rich and clever;
The golden and the goodly house, where things grow worse for ever;
There are things you need not know of, though you live and die in vain,
There are souls more sick of pleasure than you are sick of pain;
There is a game of April Fool that's played behind its door,
Where the fool remains for ever and the April comes no more,
Where the splendour of the daylight grows drearier than the dark,
And life droops like a vulture that once was such a lark:
And that is the Blue Devil that once was the Blue Bird;
For the Devil is a gentleman, and doesn't keep his word.

235 *The Song against Grocers*

GOD made the wicked Grocer
For a mystery and a sign,
That men might shun the awful shops
And go to inns to dine;

Where the bacon's on the rafter
And the wine is in the wood,
And God that made good laughter
Has seen that they are good.

The evil-hearted Grocer
Would call his mother 'Ma'am,'
And bow at her and bob at her,
Her aged soul to damn,
And rub his horrid hands and ask
What article was next,
Though *mortis in articulo*
Should be her proper text.

His props are not his children,
But pert lads underpaid,
Who call out 'Cash!' and bang about
To work his wicked trade;
He keeps a lady in a cage
Most cruelly all day,
And makes her count and calls her 'Miss'
Until she fades away.

The righteous minds of innkeepers
Induce them now and then
To crack a bottle with a friend
Or treat unmoneyed men,
But who hath seen the Grocer
Treat housemaids to his teas
Or crack a bottle of fish sauce
Or stand a man a cheese?

He sells us sands of Araby
As sugar for cash down;
He sweeps his shop and sells the dust
The purest salt in town,
He crams with cans of poisoned meat
Poor subjects of the King,
And when they die by thousands
Why, he laughs like anything.

The wicked Grocer groces
In spirits and in wine,
Not frankly and in fellowship
As men in inns do dine;
But packed with soap and sardines
And carried off by grooms,
For to be snatched by Duchesses
And drunk in dressing-rooms.

The hell-instructed Grocer
Has a temple made of tin,
And the ruin of good innkeepers
Is loudly urged therein;
But now the sands are running out
From sugar of a sort,
The Grocer trembles; for his time,
Just like his weight, is short.

ROBERT FROST

1874–1963

236 *The Objection to Being Stepped On*

At the end of the row
I stepped on the toe
Of an unemployed hoe.
It rose in offense
And struck me a blow
In the seat of my sense.
It wasn't to blame
But I called it a name.
And I must say it dealt
Me a blow that I felt
Like malice prepense.
You may call me a fool,
But *was* there a rule
The weapon should be
Turned into a tool?
And what do we see?
The first tool I step on
Turned into a weapon.

237 *Lucretius versus the Lake Poets*

'Nature I loved; and next to Nature, Art.'

DEAN, adult education may seem silly.
What of it, though? I got some willy-nilly
The other evening at your college deanery.
And grateful for it (let's not be facetious!)
For I thought Epicurus and Lucretius
By Nature meant the Whole Goddam Machinery.
But you say that in college nomenclature
The only meaning possible for Nature
In Landor's quatrain would be Pretty Scenery.
Which makes opposing it to Art absurd
I grant you—if you're sure about the word.
God bless the Dean and make his deanship plenary.

238 *Clear and Colder*

WIND, the season-climate mixer,
In my Witches' Weather Primer
Says, to make this Fall Elixir
First you let the summer simmer,
Using neither spoon nor skimmer,

Till about the right consistence.
(This like fate by stars is reckoned,
None remaining in existence
Under magnitude the second.)

Then take some leftover winter
Far to north of the St Lawrence.
Leaves to strip and branches splinter,
Bring on wind. Bring rain in torrents—
Colder than the season warrants.

Dash it with some snow for powder.
If this seems like witchcraft rather,
If this seems a witches' chowder
(All my eye and Cotton Mather!),

Wait and watch the liquor settle.
I could stand whole dayfuls of it.
Wind she brews a heady kettle.
Human beings love it—love it.
Gods above are not above it.

239

An Answer

But Islands of the Blessèd, bless you, son,
I never came upon a blessed one.

DON MARQUIS

1874–1937

240

from *archy and mehitabel*

(archy is a cockroach; mehitabel is a cat; archy records their sayings and doings on the boss's typewriter late at night)

(i)

[*the song of mehitabel*]

this is the song of mehitabel
of mehitabel the alley cat
as i wrote you before boss
mehitabel is a believer
in the pythagorean
theory of the transmigration
of the soul and she claims
that formerly her spirit
was incarnated in the body
of cleopatra
that was a long time ago
and one must not be
surprised if mehitabel
has forgotten some of her
more regal manners

i have had my ups and downs
but wotthehell wotthehell
yesterday sceptres and crowns
fried oysters and velvet gowns
and today i herd with bums
but wotthehell wotthehell
i wake the world from sleep
as i caper and sing and leap
when i sing my wild free tune
wotthehell wotthehell
under the blear eyed moon
i am pelted with cast off shoon
but wotthehell wotthehell

do you think that i would change
my present freedom to range
for a castle or moated grange
wotthehell wotthehell
cage me and i d go frantic
my life is so romantic
capricious and corybantic
and i m toujours gai toujours gai

i know that i am bound
for a journey down the sound
in the midst of a refuse mound
but wotthehell wotthehell
oh i should worry and fret
death and i will coquette
there s a dance in the old dame yet
toujours gai toujours gai

i once was an innocent kit
wotthehell wotthehell
with a ribbon my neck to fit
and bells tied onto it
o wotthehell wotthehell
but a maltese cat came by
with a come hither look in his eye
and a song that soared to the sky
and wotthehell wotthehell
and i followed adown the street
the pad of his rhythmical feet
o permit me again to repeat

wotthehell wotthehell
my youth i shall never forget
but there s nothing i really regret
wotthehell wotthehell
there s a dance in the old dame yet
toujours gai toujours gai

the things that i had not ought to
i do because i ve gotto
wotthehell wotthehell
and i end with my favorite motto
toujours gai toujours gai

boss sometimes i think
that our friend mehitabel
is a trifle too gay

(ii)

[*archy interviews a pharaoh*]

boss i went
and interviewed the mummy
of the egyptian pharaoh
in the metropolitan museum
as you bade me to do

what ho
my regal leatherface
says i

greetings
little scatter footed
scarab
says he

kingly has been
says i
what was your ambition
when you had any

insignificant
and journalistic insect
says the royal crackling
in my tender prime
i was too dignified

to have anything as vulgar
as ambition
the ra ra boys
in the seti set
were too haughty
to be ambitious
we used to spend our time
feeding the ibises
and ordering
pyramids sent home to try on
but if i had my life
to live over again
i would give dignity
the regal razz
and hire myself out
to work in a brewery

old tan and tarry
says i
i detect in your speech
the overtones
of melancholy

yes i am sad
says the majestic mackerel
i am as sad
as the song
of a soudanese jackal
who is wailing for the blood red
moon he cannot reach and rip

on what are you brooding
with such a wistful
wishfulness
there in the silences
confide in me
my imperial pretzel
says i

i brood on beer
my scampering whiffle snoot
on beer says he

my sympathies
are with your royal
dryness says i

my little pest
says he
you must be respectful
in the presence
of a mighty desolation
little archy
forty centuries of thirst
look down upon you
oh by isis
and by osiris
says the princely raisin
and by pish and phthush and phthah
by the sacred book perembru
and all the gods
that rule from the upper
cataract of the nile
to the delta of the duodenum
i am dry
i am as dry
as the next morning mouth
of a dissipated desert
as dry as the hoofs
of the camels of timbuctoo
little fussy face
i am as dry as the heart
of a sand storm
at high noon in hell
i have been lying here
and there
for four thousand years
with silicon in my esophagus
and gravel in my gizzard
thinking
thinking
thinking
of beer

divine drouth
says i
imperial fritter

continue to think
there is no law against
that in this country
old salt codfish
if you keep quiet about it
not yet

what country is this
asks the poor prune

my reverend juicelessness
this is a beerless country
says i

well well said the royal
desiccation
my political opponents back home
always maintained
that i would wind up in hell
and it seems they had the right dope

and with these hopeless words
the unfortunate residuum
gave a great cough of despair
and turned to dust and debris
right in my face
it being the only time
i ever actually saw anybody
put the cough
into sarcophagus

dear boss as i scurry about
i hear of a great many
tragedies in our midsts
personally i yearn
for some dear friend to pass over
and leave to me
a boot legacy
yours for the second coming
of gambrinus

archy

Gambrinus] mythical king credited with the first brewing of beer

HARRY GRAHAM

1874–1936

241

L'Enfant Glacé

WHEN Baby's cries grew hard to bear
I popped him in the Frigidaire.
I never would have done so if
I'd known that he'd be frozen stiff.
My wife said: 'George, I'm so unhappé!
Our darling's now completely *frappé*!'

242

Waste

I HAD written to Aunt Maud,
Who was on a trip abroad,
When I heard she'd died of cramp
Just too late to save the stamp.

243

Opportunity

WHEN Mrs Gorm (Aunt Eloise)
Was stung to death by savage bees,
Her husband (Prebendary Gorm)
Put on his veil, and took the swarm.
He's publishing a book next May
On 'How to Make Bee-keeping Pay'.

244

Grandpapa

THIS is a portrait. Here one can
  Descry those purely human features
Whereby, since first the world began,
Man has with ease distinguished Man
  From humbler fellow-creatures
And seldom, whatsoe'er his shape,
Mistaken him for Dog or Ape.

Inspect this subject well, and note
  The whiskers centrally divided,
The silken stock about his throat,
The loose but elegant frock-coat,
  The boots (elastic-sided),
And you'll at once remark: 'Ah, ha!
'This must, of course, be Grandpapa!'

'Tis he, of feudal types the last,
  By all his peers revered, respected;
His lines in pleasant places cast
Where churls saluted as he pass'd
  And maidens genuflected,
And, if he chanced to meet the Vicar,
The latter's pulse would beat the quicker.

As yet upon his vast estates
  No labour troubles had arisen;
There were no beggars at his gates—
He and his brother-magistrates
  Had sent them all to prison,
Knowing 'twas wiser to avoid
Encouraging the Unemployed.

Though tender-hearted, I declare,
  And often moved to righteous rages
When told that his own workmen were
Reduced to vegetarian fare
  By their starvation wages,
Such gloomy topics he'd dismiss—
He knew there was no cure for this.

Suppose some tenant, old and bent,
  By age or penury afflicted,
Failed to produce his quarter's rent,
What agony of mind it meant
  To have the man evicted!
To watch each bankrupt friend depart
Would nearly break poor Grandpa's heart.

In politics it was his rule
  To be broadminded but despotic;
In argument he kept quite cool,
Knowing a man to be a fool
  And most unpatriotic
Who differed from the views that he
Had cherished from the age of three.

I well remember, as a child,
  How much his moods perplexed and awed me;
At times irate, at others mild,
Alternately he frowned and smiled,
  Would censure or applaud me,
And either pat me on the head
Or send me screaming off to bed.

Once I recall—a sad affair—
  When, as a child of years still tender,
I chanced to sit in *his* armchair,
He seized me roughly by the hair
  And flung me in the fender.
He had such quaint impulsive ways;
I didn't sit again for days.

Dear Grandpapa—I see him yet,
  My friend, philosopher and guide, too,
A personality, once met,
One could not possibly forget,
  Though lots of people tried to—
Founder of a distinguished line,
And worthy ancestor of mine!

CLARENCE DAY

1874–1936

245

from *Scenes from the Mesozoic*

(i)

YESTERDAY explorers found
In a cold pre-glacial mound
This pathetic little rhyme
Written in our planet's prime:

'A person of sobriety
Who cherishes propriety
Feels conscious of satiety
When in a clown's society.'

Even in this scientific
Rendering of the hieroglyphic,
We can hear a quiet sigh
Coming down from days gone by.

(ii)

ON Lido's shore
And Bailey's Beach
Beasts used to roar
Or weirdly screech.
With hearty passion
They would sport,
Where wealth and fashion
Now resort.

(iii)

WITH scents and sounds and sudden views
Of murders, rapes, or play,
The jungle was alive with news,
As cities are today.
In every age the animals
Have peacefully perused
The 'Daily Breeze' and read of pals
A-being ate or bruised.

(iv)

THE real objection to the nude,
  Apart of course from chill,
Is that it looks a trifle crude;
  It doesn't fill the bill.

The fact is we need some disguise,
At least in other people's eyes;
Though not so very long ago,
Tradition says, this wasn't so.

The well-dressed beast in ages gone
Had absolutely nothing on,
And yet—if one of the elect—
Achieved with ease the right effect.

(v)

IN mesozoic days a war,
However short, went quite as far
As those du Pont equips, or Krupp.
Like us, they ate each other up.

(vi)

SOMETIMES a beast would learn to brood
And cultivate an awesome mood,
Or moan in quite a striking way.
This brought him fame. It does today.

(vii)

IN those old days they tell us of,
  In many a bad old land,
Supplies, though scant, of mother love
  Exceeded the demand.
No mother ever gave her all:
  The very thought would gall her,
But even when her love was small,
  The call for it was smaller.

(viii)

ONCE, they say, a bat-like brute,
Which began to evolute
Long before the apes or others,
Grew so man-like he was hated
And at length annihilated
    By his brothers;

(ix)

AND a shark that once began,
By mistake, to be a man,
Finding nobody could bear him,
Prayed to God Himself to spare him.

But the apes—though not so vicious
To begin with—were ambitious.

ROBERT W. SERVICE

1874–1958

246 *The Cremation of Sam McGee*

There are strange things done in the midnight sun
By the men who moil for gold;
The Arctic trails have their secret tales
That would make your blood run cold;
The Northern Lights have seen queer sights,
But the queerest they ever did see
Was that night on the marge of Lake Lebarge
I cremated Sam McGee.

NOW Sam McGee was from Tennessee, where the cotton blooms and blows.
Why he left his home in the South to roam 'round the Pole, God only knows.
He was always cold, but the land of gold seemed to hold him like a spell;
Though he'd often say in his homely way that 'he'd sooner live in hell.'

On a Christmas Day we were mushing our way over the Dawson trail.
Talk of your cold! through the parka's fold it stabbed like a driven nail.
If our eyes we'd close, then the lashes froze till sometimes we couldn't see;
It wasn't much fun, but the only one to whimper was Sam McGee.

And that very night, as we lay packed tight in our robes beneath the snow,
And the dogs were fed, and the stars o'erhead were dancing heel and toe,
He turned to me, and 'Cap', says he, 'I'll cash in this trip, I guess;
And if I do, I'm asking that you won't refuse my last request.'

Well, he seemed so low that I couldn't say no; then he says with a sort of moan:
'It's the cursed cold, and it's got right hold till I'm chilled clean through to the bone.
Yet 'tain't being dead—it's my awful dread of the icy grave that pains;
So I want you to swear that, foul or fair, you'll cremate my last remains.'

A pal's last need is a thing to heed, so I swore I would not fail;
And we started on at the streak of dawn; but God! he looked ghastly pale.
He crouched on the sleigh, and he raved all day of his home in Tennessee;
And before nightfall a corpse was all that was left of Sam McGee.

There wasn't a breath in that land of death, and I hurried, horror-driven,
With a corpse half hid that I couldn't get rid, because of a promise given;
It was lashed to the sleigh, and it seemed to say: 'You may tax your brawn and brains,
But you promised true, and it's up to you to cremate those last remains.'

Now a promise made is a debt unpaid, and the trail has its own stern code.
In the days to come, though my lips were dumb, in my heart how I cursed that load.
In the long, long night, by the lone firelight, while the huskies, round in a ring,
Howled out their woes to the homeless snows—O God! how I loathed the thing.

And every day that quiet clay seemed to heavy and heavier grow;
And on I went, though the dogs were spent and the grub was getting low;
The trail was bad, and I felt half mad, but I swore I would not give in;
And I'd often sing to the hateful thing, and it hearkened with a grin.

Till I came to the marge of Lake Lebarge, and a derelict there lay;
It was jammed in the ice, but I saw in a trice it was called the 'Alice May'.
And I looked at it, and I thought a bit, and I looked at my frozen chum;
Then 'Here', said I, with a sudden cry, 'is my cre-ma-tor-eum.'

Some planks I tore from the cabin floor, and I lit the boiler fire;
Some coal I found that was lying around, and I heaped the fuel higher;
The flames just soared, and the furnace roared—such a blaze you seldom see;
And I burrowed a hole in the glowing coal, and I stuffed in Sam McGee.

Then I made a hike, for I didn't like to hear him sizzle so;
And the heavens scowled, and the huskies howled, and the wind began to blow.
It was icy cold, but the hot sweat rolled down my cheeks, and I don't know why;
And the greasy smoke in an inky cloak went streaking down the sky.

I do not know how long in the snow I wrestled with grisly fear;
But the stars came out and they danced about ere again I ventured near;
I was sick with dread, but I bravely said: 'I'll just take a peep inside.
I guess he's cooked, and it's time I looked'; ... then the door I opened wide.

And there sat Sam, looking cool and calm, in the heart of the furnace roar;
And he wore a smile you could see a mile, and he said: 'Please close that door.
It's fine in here, but I greatly fear you'll let in the cold and storm—
Since I left Plumtree, down in Tennessee, it's the first time I've been warm.'

There are strange things done in the midnight sun
By the men who moil for gold;
The Arctic trails have their secret tales
That would make your blood run cold;
The Northern Lights have seen queer sights,
But the queerest they ever did see
Was that night on the marge of Lake Lebarge
I cremated Sam McGee.

E. C. BENTLEY

1875–1956

247 *Clerihews*

(i)

WYNKYN DE WORDE
Had as funny a name as ever I heard.
Of what could they have been thinking
When they called him Wynkyn?

(ii)

AFTER dinner, Erasmus
Told Colet not to be 'blas'mous'
Which Colet, with some heat,
Requested him to repeat.

(iii)

WHEN their lordships asked Bacon
How many bribes he had taken
He had at least the grace
To get very red in the face.

(iv)

HOW vigilant was Spenser
As a literary censor!
He pointed out that there were too few *Es*
In Lyly's *Euphues*.

(v)

THE intrepid Ricardo,
With characteristic bravado,
Alluded openly to Rent
Wherever he went.

(vi)

'NO, sir,' said General Sherman,
'I did *not* enjoy the sermon;
Nor I didn't git any
Kick outer the Litany.'

(vii)

'Susaddah!' exclaimed Ibsen,
'By dose is turdig cribson!
I'd better dot kiss you,
Atishoo! Atishoo!'

(viii)

'Dinner-time?' said Gilbert White,
'Yes, yes—certainly—all right.
Just let me finish this note
About the Lesser White-bellied Stoat.'

T. W. CONNOR

d. 1936

248

She Was One of the Early Birds

SHE was a dear little dicky bird,
'Chip, chip, chip,' she went,
Sweetly she sang to me
Till all my money was spent;
Then she went off song—
We parted on fighting terms,
She was one of the early birds,
And I was one of the worms.

R. P. WESTON (1878–1936) and BERT LEE (1880–1945)

249

[*Epitaph*]

BENEATH this stone lies William Burke,
A decent man entirely.
The stone was bought in a second-hand shop,
And his name wasn't Burke, it was Reilly.

E. V. KNOX

1881–1971

250

The Director

THEY made me a director,
I dreamt it in a dream;
I was a print collector
And owned a salmon stream.

They made me a director
  Of companies one or two;
I did not fear the spectre
  Of Nemesis—would you?

They made me a director
  Of companies two or three;
I bought myself a sector
  Of Sussex, near the sea.

They made me a director
  Of companies three or four;
I had a man named Hector
  To answer the front-door.

They made me a director
  Of companies four or five;
The beams of my reflector
  Lit up the laurelled drive.

They made me a director
  Of companies five or six;
I was a stern protector
  Of meal-fed pheasant chicks.

They made me a director
  Of companies six or seven
No shareholding objector
  Opposed my path to heaven.

They made me a director
  Of companies seven or eight;
The income-tax collector
  Knelt down before my gate.

They made me a director
  Of companies eight or nine;
I drank the golden nectar
  And had no other wine.

They made me a director
  Of companies nine or ten—

* * * *

'*Hullo, police-inspector!*
  Good morning, plain-clothes men!'

P. G. WODEHOUSE

1881–1975

251

Printer's Error

As o'er my latest book I pored,
   Enjoying it immensely,
I suddenly exclaimed 'Good Lord!'
   And gripped the volume tensely.
'Golly!' I cried. I writhed in pain.
'They've done it on me once again!'
   And furrows creased my brow.
I'd written (which I thought quite good)
'Ruth, ripening into womanhood,
Was now a girl who knocked men flat
And frequently got whistled at',
And some vile, careless, casual gook
Had spoiled the best thing in the book
   By printing 'not'
   (Yes, 'not', great Scott!)
   When I had written 'now'.

On murder in the first degree
   The Law, I knew, is rigid:
Its attitude, if A kills B,
   To A is always frigid.
It counts it not a trivial slip
If on behalf of authorship
You liquidate compositors.
This kind of conduct it abhors
   And seldom will allow.
Nevertheless, I deemed it best
And in the public interest
To buy a gun, to oil it well,
Inserting what is called a shell,
   And go and pot
   With sudden shot
   This printer who had printed 'not'
   When I had written 'now'.

I tracked the bounder to his den
  Through private information:
I said, 'Good afternoon', and then
  Explained the situation:
'I'm not a fussy man,' I said.
'I smile when you put "rid" for "red"
And "bad" for "bed" and "hoad" for "head"
  And "bolge" instead of "bough".
When "wone" appears in lieu of "wine"
Or if you alter "Cohn" to "Schine",
  I never make a row.
I know how easy errors are.
But this time you have gone too far
By printing "not" when you knew what
  I really wrote was "now".
Prepare,' I said, 'to meet your God
Or, as you'd say, your Goo or Bod,
  Or possibly your Gow.'

A few weeks later into court
  I came to stand my trial.
The Judge was quite a decent sort.
  He said, 'Well, cocky, I'll
Be passing sentence in a jiff,
And so, my poor unhappy stiff,
If you have anything to say,
Now is the moment. Fire away.
  You have?'
  I said, 'And how!
Me lud, the facts I don't dispute.
I did, I own it freely, shoot
This printer through the collar stud.
What else could I have done, me lud?
  He'd printed "not" ...'
  The judge said, '*What!*
  When you had written "now"?
God bless my soul! Gadzooks!' said he.
'The blighters did that once to me.
  A dirty trick, I trow.
I hereby quash and override
The jury's verdict. Gosh!' he cried.
'Give me your hand. Yes, I insist,
You splendid fellow! Case dismissed.'
  (Cheers, and a Voice 'Wow-wow!')

A statue stands against the sky,
  Lifelike and rather pretty.
'Twas recently erected by
  The P.E.N. committee.
And many a passer-by is stirred,
For on the plinth, if that's the word,
In golden letters you may read
'This is the man who did the deed.
  His hand set to the plough,
He did not sheathe the sword, but got
A gun at great expense and shot
The human blot who'd printed "not"
  When he had written "now".
He acted with no thought of self,
Not for advancement, not for pelf,
But just because it made him hot
To think the man had printed "not"
  When he had written "now".'

FRANKLIN P. ADAMS ('F. P. A.')

1881–1960

252

Those Two Boys

WHEN Bill was a lad he was terribly bad.
  He worried his parents a lot;
He'd lie and he'd swear and pull little girls' hair;
  His boyhood was naught but a blot.

At play and in school he would fracture each rule—
  In mischief from autumn to spring;
And the villagers knew when to manhood he grew
  He would never amount to a thing.

When Jim was a child he was not very wild;
  He was known as a good little boy;
He was honest and bright and the teacher's delight—
  To his mother and father a joy.

All the neighbors were sure that his virtue'd endure,
  That his life would be free of a spot;
They were certain that Jim had a great head on him
  And that Jim would amount to a lot.

And Jim grew to manhood and honor and fame
  And bears a good name;
While Bill is shut up in a dark prison cell—
  You never can tell.

WILLIAM HARGREAVES

1881–1941

253 *Burlington Bertie from Bow*

I'M Bert, p'rhaps you've heard of me, Bert, you've had word of me,
Jogging along, hearty and strong, living on plates of fresh air:
I dress up in fashion, and, when I am feeling depressed,
I shave from my cuff all the whiskers and fluff, stick my hat on and
  toddle up West.

I'm Burlington Bertie, I rise at ten-thirty, and saunter along like a toff,
I walk down the Strand with my gloves on my hand and I walk back
  again with them off.
I'm all airs and graces, correct easy paces, without food so long I've
  forgot where my face is.
I'm Bert, Bert, I haven't a shirt, but my people are well off, you know!
Nearly everyone knows me, from Smith to Lord Rosebery,
I'm Burlington Bertie from Bow!

I stroll with Lord Hurlington, roll in the Burlington,
Call for champagne, walk out again, come back and borrow the ink.
I live most expensive—like Tom Lipton I'm in the swim:
He's got so much 'oof' that he sleeps on the roof, and I live in the room
  over him.

I'm Burlington Bertie, I rise at ten-thirty, then saunter along Temple Bar,
As round there I skip, I keep shouting 'Pip! Pip!' and the darned fools
  think I'm in my car.

At Rothschild's I swank it, my body I plank it right on his front door with the 'Mail' for a blanket,
I'm Bert, Bert, and Rothschild was hurt; he said 'You can't sleep there.' I said 'Oh?'
He said '*I'm* Rothschild, sonny!' I said 'That's damned funny,
I'm Burlington Bertie from Bow!'

My pose, tho' ironical, shows that my monocle
Holds up my face, keeps it in place, stops it from slipping away.
Cigars—I smoke thousands, I usually deal in the Strand,
But you've got to take care when you're getting them there, or some idiot might stand on your hand.

I'm Burlington Bertie, I rise at ten-thirty, then Buckingham Palace I view;
I stand in the yard while they're changing the guard and the King shouts across 'Toodle-oo.'
The Prince of Wales' brother, along with some other, slaps me on the back and says, 'Come and see Mother.'
I'm Bert, Bert, and Royalty's hurt, when they ask me dine, I say 'No,
I've just had a banana with Lady Diana,
I'm Burlington Bertie from Bow!'

BILLY MERSON

1881–1947

254 *The Spaniard that Blighted My Life*

LIST to me while I tell you
Of the Spaniard that blighted my life;
List to me while I tell you
Of the man who pinched my future wife.
'Twas at the bull-fight where we met him,
We'd been watching his daring display,
And while I'd gone out for some nuts and a programme
The dirty dog stole her away.
Oh yes! oh yes!
But I've sworn that I'll have my revenge!

If I catch Alphonso Spagoni the Toreador,
With one mighty swipe I will dislocate his bally jaw!
I'll find this bull-fighter, I will,
And when I catch the bounder, the blighter I'll kill.
He shall die! he shall die! he shall die tiddly-i-ti-ti-ti-ti-ti-ti!
He shall die! he shall die!
For I'll raise a bunion on his Spanish onion
If I catch him bending tonight!

Yes, when I catch Spagoni
He will wish that he'd never been born;
And for this special reason,
My stiletto I've fetched out of pawn.
It cost me five shillings to fetch it,
This expense it has caused me much pain,
But the pawnbroker's promised when I've killed Spagoni
He'll take it in pawn once again.
Oh yes! oh, yes!
So tonight there will be dirty work.

If I catch Alphonso Spagoni, the Toreador,
With one mighty swipe I will dislocate his bally jaw!
I'll find this bull-fighter, I will,
And when I catch the bounder, the blighter I'll kill.
He shall die! he shall die! he shall die tiddly-i-ti-ti-ti-ti-ti-ti!
He shall die! he shall die!
For I'll raise a bunion on his Spanish onion
If I catch him bending tonight!

JAMES JOYCE

1882–1941

255 *Post Ulixem Scriptum*

(*Air*: 'Molly Brannigan')

MAN dear, did you never hear of buxom Molly Bloom at all,
As plump an Irish beauty, Sir, as any Levi-Blumenthal?
If she sat in the viceregal box Tim Healy'd have no room at all,
    But curl up in a corner at a glance from her eye.

Post Ulixem Scriptum] after the writing of *Ulysses*   *Tim Healy*] first Governor-General of the Irish Free State

The tale of her ups and downs would aisy fill a handybook
That would cover the two worlds at once from Gibraltar 'cross to Sandy
        Hook.
But now that tale is told, ochone, I've lost my daring dandy look:
    Since Molly Bloom has left me here alone for to cry.

Man dear, I remember when my roving time was troubling me
We picknicked fine in storm or shine in France and Spain and Hungary
And she said I'd be her first and last while the wine I poured went
        bubbling free
    Now every male you meet with has a finger in her pie.
Man dear, I remember with all the heart and brain of me
I arrayed her for the bridal but, O, she proved the bane of me.
With more puppies sniffing round her than the wooers of Penelope
    She's left me on her doorstep like a dog for to die.

My left eye is wake and his neighbour full of water, man.
I cannot see the lass I limned as Ireland's gamest Daughter, man,
When I hear her lovers tumbling in their thousands for to court her,
        man,
    If I was sure I'd not be seen I'd sit down and cry.
May you live, may you love like this gaily spinning earth of ours,
And every morn a gallant sun awake you with new wealth of gold
But if I cling like a child to the clouds that are your petticoats
    O Molly, handsome Molly, sure you won't let me die!

256 [*A Blurb for* Anna Livia Plurabelle]

Buy a book in brown paper
From Faber and Faber
To see Annie Liffey trip, tumble and caper.
Sevensinns in her singthings,
Plurabells on her prose,
Seashell ebb music wayriver she flows.

257 [*A Blurb for* Haveth Childers Everywhere]

Humptydump Dublin squeaks through his norse,
Humptydump Dublin hath a horriple vorse,
        And, with all his kinks english
        Plus his irishmanx brogues,
Humptydump Dublin's grandada of rogues.

A. A. MILNE

1882–1956

258 *The King's Breakfast*

THE King asked
The Queen, and
The Queen asked
The Dairymaid:
'Could we have some butter for
The Royal slice of bread?'
The Queen asked
The Dairymaid,
The Dairymaid
Said: 'Certainly,
I'll go and tell
The cow
Now
Before she goes to bed.'

The Dairymaid
She curtsied,
And went and told
The Alderney:
'Don't forget the butter for
The Royal slice of bread.'
The Alderney
Said sleepily:
'You'd better tell
His Majesty
That many people nowadays
Like marmalade
Instead.'

The Dairymaid
Said: 'Fancy!'
And went to
Her Majesty.
She curtsied to the Queen, and
She turned a little red:
'Excuse me,
Your Majesty,

For taking of
The liberty,
But marmalade is tasty, if
It's very
Thickly
Spread.'

The Queen said:
'Oh!'
And went to
His Majesty:
'Talking of the butter for
The Royal slice of bread,
Many people
Think that
Marmalade
Is nicer.
Would you like to try a little
Marmalade
Instead?'

The King said:
'Bother!'
And then he said:
'Oh, deary me!'
The King sobbed: 'Oh, deary me!'
And went back to bed.
'Nobody,'
He whimpered,
'Could call me
A fussy man;
I *only* want
A little bit
Of butter for
My bread!'

The Queen said:
'There, there!'
And went to
The Dairymaid.
The Dairymaid
Said: 'There, there!'
And went to the shed.
The cow said:

'There, there!
I didn't really
Mean it;
Here's milk for his porringer
And butter for his bread.'

The Queen took
The butter
And brought it to
His Majesty;
The King said:
'Butter, eh?'
And bounced out of bed.
'Nobody,' he said:
As he kissed her
Tenderly,
'Nobody,' he said,
As he slid down
The banisters,
'Nobody,
My darling,
Could call me
A fussy man—
BUT
I do like a little bit of butter to my bread!'

JAMES STEPHENS

1882–1950

259

A Glass of Beer

THE lanky hank of a she in the inn over there
Nearly killed me for asking the loan of a glass of beer;
May the devil grip the whey-faced slut by the hair,
And beat bad manners out of her skin for a year.

That parboiled ape, with the toughest jaw you will see
On virtue's path, and a voice that would rasp the dead,
Came roaring and raging the minute she looked at me,
And threw me out of the house on the back of my head!

If I asked her master he'd give me a cask a day;
But she, with the beer at hand, not a gill would arrange!
May she marry a ghost and bear him a kitten, and may
The High King of Glory permit her to get the mange.

260 *Blue Blood*

WE thought at first, this man is a king for sure,
Or the branch of a mighty and ancient and famous lineage
—That silly, sulky, illiterate, black-avised boor
Who was hatched by foreign vulgarity under a hedge!

The good men of Clare were drinking his health in a flood,
And gazing, with me, in awe at the princely lad;
And asking each other from what bluest blueness of blood
His daddy was squeezed, and the pa of the da of his dad?

We waited there, gaping and wondering, anxiously,
Until he'd stop eating, and let the glad tidings out;
And the slack-jawed booby proved to the hilt that he
Was lout, son of lout, by old lout, and was da to a lout!

CLEMENT ATTLEE

1883–1967

261 [*On his Own Career*]

FEW thought he was even a starter,
There were many who thought themselves smarter,
    But he ended PM,
    CH and OM,
An earl and a knight of the garter.

J. C. SQUIRE

1884–1958

262

The Dilemma

GOD heard the embattled nations sing and shout
'Gott strafe England!' and 'God save the King!'
God this, God that, and God the other thing—
'Good God!' said God, 'I've got my work cut out.'

KEITH PRESTON

1884–1927

263

Lapsus Linguae

WE wanted Li Wing
  But we winged Willie Wong.
A sad but excusable
  Slip of the tong.

D. H. LAWRENCE

1885–1930

264

I Am in a Novel—

I READ a novel by a friend of mine
in which one of the characters was me,
the novel it sure was mighty fine
but the funniest thing that could be

was me, or what was supposed for me,
for I had to recognise
a few of the touches, like a low-born jake,
but the rest was a real surprise.

Well damn my eyes! I said to myself.
Well damn my little eyes!
If this is what Archibald thinks I am
he sure thinks a lot of lies.

Well think o' that now, think o' that!
That's what he sees in me!
I'm about as much like a Persian cat,
or a dog with a harrowing flea.

My Lord! a man's friends' ideas of him
would stock a menagerie
with a marvellous outfit! How did Archie see
such a funny pup in me?

265 *When I Read Shakespeare—*

When I read Shakespeare I am struck with wonder
that such trivial people should muse and thunder
in such lovely language.

Lear, the old buffer, you wonder his daughters
didn't treat him rougher,
the old chough, the old chuffer!

And Hamlet, how boring, how boring to live with,
so mean and self-conscious, blowing and snoring
his wonderful speeches, full of other folks' whoring!

And Macbeth and his Lady, who should have been choring,
such suburban ambition, so messily goring
old Duncan with daggers!

How boring, how small Shakespeare's people are!
Yet the language so lovely! like the dyes from gas-tar.

266 *Innocent England*

Oh what a pity, Oh! don't you agree
that figs aren't found in the land of the free!

Fig-trees don't grow in my native land;
there's never a fig-leaf near at hand

when you want one; so I did without;
and that is what the row's about.

Virginal, pure policemen came
and hid their faces for very shame,

while they carried the shameless things away
to gaol, to be hid from the light of day.

And Mr Mead, that old, old lily
said: 'Gross! coarse! hideous!'—and I, like a silly

thought he meant the faces of the police-court officials,
and how right he was, and I signed my initials

to confirm what he said; but alas, he meant
my pictures, and on the proceedings went.

The upshot was, my pictures must burn
that English artists might finally learn

when they painted a nude, to put a *cache sexe* on,
a cache sexe, a cache sexe, or else begone!

A fig-leaf; or, if you cannot find it
a wreath of mist, with nothing behind it.

A wreath of mist is the usual thing
in the north, to hide where the turtles sing.

Though they never sing, they never sing,
don't you dare to suggest such a thing

or Mr Mead will be after you.
—But what a pity I never knew

A wreath of English mist would do
as a cache sexe! I'd have put a whole fog.

But once and forever barks the old dog,
so my pictures are in prison, instead of in the Zoo.

Mr Mead] the London magistrate before whom the case of Lawrence's pictures was brought in August 1929

EZRA POUND

1885–1972

267 *Soirée*

UPON learning that the mother wrote verses,
And that the father wrote verses,
And that the youngest son was in a publisher's office,
And that the friend of the second daughter was undergoing a novel,
The young American pilgrim
Exclaimed:
    'This is a darn'd clever bunch!'

268 *Ancient Music*

WINTER is icummen in,
Lhude sing Goddamm,
Raineth drop and staineth slop,
And how the wind doth ramm!
            Sing: Goddamm.
Skiddeth bus and sloppeth us,
An ague hath my ham.
Freezeth river, turneth liver,
            Damn you, sing: Goddamm.
Goddamm, Goddamm, 'tis why I am, Goddamm.
            So 'gainst the winter's balm.
Sing goddamm, damm, sing Goddamn,
Sing goddamm, sing goddamm, DAMM.

269 *The Temperaments*

NINE adulteries, 12 liaisons, 64 fornications and something approaching
  a rape
Rest nightly upon the soul of our delicate friend Florialis,
And yet the man is so quiet and reserved in demeanour
That he passes for both bloodless and sexless.
Bastidides, on the contrary, who both talks and writes of nothing save
  copulation,
Has become the father of twins,
But he accomplished this feat at some cost;
He had to be four times cuckold.

270 *Les Millwin*

THE little Millwins attend the Russian Ballet.
The mauve and greenish souls of the little Millwins
Were seen lying along the upper seats
Like so many unused boas.

The turbulent and undisciplined host of art students—
The rigorous deputation from 'Slade'—
Was before them.

With arms exalted, with fore-arms
Crossed in great futuristic X's, the art students
Exulted, they beheld the splendours of *Cleopatra*.

And the little Millwins beheld these things;
With their large and anæmic eyes they looked out upon this configuration.

Let us therefore mention the fact,
For it seems to us worthy of record.

271 *The Lake Isle*

O GOD, O Venus, O Mercury, patron of thieves,
Give me in due time, I beseech you, a little tobacco-shop,
With the little bright boxes
    piled up neatly upon the shelves
And the loose fragrant cavendish
    and the shag,
And the bright Virginia
    loose under the bright glass cases,
And a pair of scales not too greasy,
And the whores dropping in for a word or two in passing,
For a flip word, and to tidy their hair a bit.

O God, O Venus, O Mercury, patron of thieves,
Lend me a little tobacco-shop,
    or install me in any profession
Save this damn'd profession of writing,
    where one needs one's brains all the time.

W. N. EWER

1885–1977

272

HOW odd
Of God
To choose
The Jews.

ANONYMOUS

273

HIS Son's
A Jew.
I thought
You knew.

HUMBERT WOLFE

1886–1940

274

YOU cannot hope
to bribe or twist,
thank God! the
British journalist.

But, seeing what
the man will do
unbribed, there's
no occasion to.

MAURICE HARE

1886–1967

275

Determinism

THERE was a young man who said, 'Damn!
It is borne in upon me I am
  An engine that moves
  In predestinate grooves,
I'm not even a bus, I'm a tram.'

276

Alfred de Musset

ALFRED DE MUSSET
Used to call his cat Pusset.
His accent was affected.
That was only to be expected.

RUPERT BROOKE

1887–1915

277

Sonnet: In Time of Revolt

THE thing must End. I am no boy! I AM
  NO BOY!! being twenty-one. Uncle, you make
  A great mistake, a very great mistake,
In chiding me for letting slip a 'Damn!'
What's more, you called me 'Mother's one ewe lamb,'
  Bade me 'refrain from swearing—for *her* sake—
  Till I'm grown up' . . . —By God! I think you take
Too much upon you, Uncle William!

You say I am your brother's only son.
I know it. And, 'What of it?' I reply.
My heart's resolvèd. *Something must be done.*
So shall I curb, so baffle, so suppress
This too avuncular officiousness,
Intolerable consanguinity.

GUS KAHN

1886–1941

278

Makin' Whoopee

EVERY time I hear that march from Lohengrin,
I am glad I'm on the outside looking in.
I have heard a lot of married people talk,
And I know that marriage is a long, long walk.
To some people weddings mean romance,
But I prefer a picnic or a dance.

Another bride, another June,
Another sunny honeymoon,
Another season, another reason,
For makin' whoopee!
A lot of shoes, a lot of rice,
The groom is nervous, he answers twice,
It's really killing that he's so willing
To make whoopee!

Picture a little love-nest,
Down where the roses cling,
Picture the same sweet love-nest,
Think what a year can bring.
He's washing dishes and baby clothes,
He's so ambitious he even sews.
But don't forget, folks,
That's what you get, folks,
For makin' whoopee!

Down through the countless ages
You'll find it everywhere,
Somebody makes good wages,
Somebody wants her share.
She calls him toodles, and rolls her eyes,
She makes him strudles, and bakes him pies,
What is it all for?
It's so he'll fall for
Makin' whoopee.

Another year, or maybe less,
What's this I hear, well can't you guess?
She feels neglected and he's suspected
Of makin' whoopee.
She sits alone most every night,
He doesn't 'phone or even write,
He says he's busy but she says 'is he?
He's makin' whoopee.'
He doesn't make much money,
Five thousand dollars per,
Some judge who thinks he's funny
Says 'You pay six to her.'
He says 'Now judge suppose I fail?'
The judge says 'Bud, right into jail,
You'd better keep her, you'll find it cheaper
Than makin' whoopee.'

EDITH SITWELL

1887–1964

279 *Sir Beelzebub*

WHEN
Sir
Beelzebub called for his syllabub in the hotel in Hell
    Where Proserpine first fell,
Blue as the gendarmerie were the waves of the sea,
    (Rocking and shocking the bar-maid).
Nobody comes to give him his rum but the
Rim of the sky hippopotamus-glum
Enhances the chances to bless with a benison
Alfred Lord Tennyson crossing the bar laid
With cold vegetation from pale deputations
Of temperance workers (all signed In Memoriam)
Hoping with glory to trip up the Laureate's feet,
    (Moving in classical metres) . . .
Like Balaclava, the lava came down from the
Roof, and the sea's blue wooden gendarmerie
Took them in charge while Beelzebub roared for his rum.
    . . . None of them come!

T. S. ELIOT

1888–1965

280 from *Five-Finger Exercises*

[*Lines for Cuscuscaraway and Mirza Murad Ali Beg*]

How unpleasant to meet Mr Eliot!
With his features of clerical cut,
And his brow so grim
And his mouth so prim
And his conversation, so nicely
Restricted to What Precisely
And If and Perhaps and But.
How unpleasant to meet Mr Eliot!
With a bobtail cur
In a coat of fur
And a porpentine cat
And a wopsical hat:
How unpleasant to meet Mr Eliot!
(Whether his mouth be open or shut).

281 *Bustopher Jones: The Cat About Town*

Bustopher Jones is *not* skin and bones—
In fact, he's remarkably fat.
He doesn't haunt pubs—he has eight or nine clubs,
For he's the St James's Street Cat!
He's the cat we all greet as he walks down the street
In his coat of fastidious black:
No commonplace mousers have such well-cut trousers
Or such an impeccable back.
In the whole of St James's the smartest of names is
The name of this Brummell of Cats;
And we're all of us proud to be nodded or bowed to
By Bustopher Jones in white spats!

His visits are occasional to the *Senior Educational*
And it is against the rules
For any one Cat to belong both to that
And the *Joint Superior Schools*.

For a similar reason, when game is in season
He is found, not at *Fox's*, but *Blimp's*;
But he's frequently seen at the gay *Stage and Screen*
Which is famous for winkles and shrimps.
In the season of venison he gives his ben'son
To the *Pothunter's* succulent bones;
And just before noon's not a moment too soon
To drop in for a drink at the *Drones.*
When he's seen in a hurry there's probably curry
At the *Siamese*—or at the *Glutton*;
If he looks full of gloom then he's lunched at the *Tomb*
On cabbage, rice pudding and mutton.

So, much in this way, passes Bustopher's day—
At one club or another he's found.
It can be no surprise that under our eyes
He has grown unmistakably round.
He's a twenty-five pounder, or I am a bounder,
And he's putting on weight every day:
But he's so well preserved because he's observed
All his life a routine, so he'd say.
Or, to put it in rhyme: 'I shall last out my time'
Is the word for this stoutest of Cats.
It must and it shall be Spring in Pall Mall
While Bustopher Jones wears white spats!

SIR GEORGE ROSTREVOR HAMILTON

1888–1967

282 *Don's Holiday*

PROFESSOR ROBINSON each summer beats
The fishing record of the world—such feats
As one would hardly credit from a lesser
Person than a history professor.

283 *To a Pessimist*

YOUR volume proves that 'Nothing is worth while.'
  Good man! I think you've covered all the ground;
And yet—and yet—that beatific smile
  With which you put your pen down? Pray expound.

RONALD KNOX

1888–1957

284 [*Exchange and Mart*]

AN Anglican curate in want
Of a second-hand portable font
    Will exchange for the same
    A photo (with frame)
Of the Bishop-Elect of Vermont.

285 [*The Modernist's Prayer*]

O GOD, forasmuch as without Thee
We are not enabled to doubt Thee,
    Help us all by Thy grace
    To convince the whole race
It knows nothing whatever about Thee.

IRVING BERLIN

1888–1989

286 *A Couple of Swells*

WE'RE a couple of swells; we stop at the best hotels.
But we prefer the country far away from the city smells.
We're a couple of sports, the pride of the tennis courts.
In June, July and August we look cute when we're dressed in shorts.

The Vanderbilts have asked us up for tea.
We don't know how to get there, no siree.

We would drive up the Avenue but we haven't got the price.
We would skate up the Avenue, but there isn't any ice.
We would ride on a bicycle, but we haven't got a bike.
So we'll walk up the Avenue.
Yes, we'll walk up the Avenue,
And to walk up the Avenue's what we like.

Wall Street Bankers are we, with plenty of currency.
We'd open up the safe, but we forgot where we put the key.
We're the favourite lads of girls in the picture ads.
We'd like to tell you who we kissed last night, but we can't be cads.

The Vanderbilts are waiting at the club.
But how are we to get there? That's the rub.

We would sail up the Avenue, but we haven't got a yacht.
We would drive up the Avenue, but the horse we had was shot.
We would ride on a trolley car, but we haven't got the fare.
So we'll walk up the Avenue.
Yes, we'll walk up the Avenue,
Yes, we'll walk up the Avenue 'till we're there.

CONRAD AIKEN

1889–1973

287 *Animula vagula blandula*

ANIMULA vagula blandula,
is it true that your origin's glandular?
Must you twang for the Lord
an umbilical chord
like all other impropagandula?

Animula vagula blandula] little wandering gentle soul

DION TITHERAGE

1889–1934

288 *And Her Mother Came Too*

I SEEM to be the victim of a cruel jest,
It dogs my footsteps with the girl I love the best.
She's just the sweetest thing that I have ever known,
But still we never get the chance to be alone.

My car will meet her—And her mother comes too!
It's a two-seater—Still her mother comes too!
At Ciro's when I am free, at dinner, supper or tea,
She loves to shimmy with me—And her mother does too!
We buy her trousseau—And her mother comes too!
Asked *not* to do so—Still her mother comes too!
She simply can't take a snub, I go and sulk at the club,
Then have a bath and a rub—And her brother comes too!

There may be times when couples need a chaperone,
But mothers ought to learn to leave a chap alone.
I wish they'd have a heart and use their common sense
For three's a crowd, and more, it's treble the expense.

We lunch at Maxim's—And her mother comes too!
How large a snack seems—When her mother comes too!
And when they're visiting me, we finish afternoon tea,
She loves to sit on my knee—And her mother does too!
To golf we started—And her mother came too!
Three bags I carted—When her mother came too!
She fainted just off the tee, my darling whisper'd to me
'Jack, dear, at last we are free!'—But her mother came to!

STANLEY HOLLOWAY

1890–1982

289

Old Sam

It occurred on the evening before Waterloo
And troops were lined up on Parade,
And Sergeant inspecting 'em, he was a terror
Of whom every man was afraid—

All excepting one man who was in the front rank,
A man by the name of Sam Small,
And 'im and the Sergeant were both 'daggers drawn',
They thought 'nowt' of each other at all.

As Sergeant walked past he was swinging his arm,
And he happened to brush against Sam,
And knocking his musket clean out of his hand
It fell to the ground with a slam.

'Pick it oop,' said Sergeant, abrupt like but cool,
But Sam with a shake of his head
Said, 'Seeing as tha' knocked it out of me hand,
P'raps tha'll pick the thing oop instead.'

'Sam, Sam, pick oop tha' musket,'
The Sergeant exclaimed with a roar.
Sam said 'Tha' knocked it doon, Reet!
Then tha'll pick it oop, or it stays where it is, on 't floor.'

The sound of high words
Very soon reached the ears of an Officer, Lieutenant Bird,
Who says to the Sergeant, 'Now what's all this 'ere?'
And the Sergeant told what had occurred.

'Sam, Sam, pick oop tha' musket,'
Lieutenant exclaimed with some heat.
Sam said 'He knocked it doon, Reet! then he'll pick it oop,
Or it stays where it is, at me feet.'

It caused quite a stir when the Captain arrived
To find out the cause of the trouble;
And every man there, all excepting Old Sam,
Was full of excitement and bubble.

'Sam, Sam, pick oop tha' musket,'
Said Captain for strictness renowned.
Sam said 'He knocked it doon, Reet!
Then he'll pick it oop, or it stays where it is on't ground.'

The same thing occurred when the Major and Colonel
Both tried to get Sam to see sense,
But when Old Duke o' Wellington came into view
Well, the excitement was tense.

Up rode the Duke on a lovely white 'orse,
To find out the cause of the bother;
He looks at the musket and then at Old Sam
And he talked to Old Sam like a brother,

'Sam, Sam, pick oop tha' musket,'
The Duke said as quiet as could be,
'Sam, Sam, pick oop tha' musket
Coom on, lad, just to please me.'

'Alright, Duke,' said Old Sam, 'just for thee I'll oblige,
And to show thee I meant no offence.'
So Sam picked it up, 'Gradeley, lad,' said the Duke,
'Right-o, boys, let battle commence.'

ANOMYMOUS

290

Spring in the Bronx

SPRING is sprung,
Duh grass is riz
I wonder where dem boidies is.

Duh little boids is on duh wing—
But dat's absoid:
Duh little wing is on duh boid.

ANONYMOUS

291 *Soldiers' Songs of the First World War*

(i)

I DON'T want to be a soldier,
I don't want to go to war.
I'd rather stay at home,
Around the streets to roam,
And live on the earnings of a —— lady-typist.
I don't want a bayonet in my belly,
I don't want my —— shot away.
I'd rather stay in England,
In merry, merry England,
And —— my bloody life away.

(ii)

WE are Fred Karno's army,
The ragtime infantry:
We cannot fight, we cannot shoot,
What earthly use are we!
And when we get to Berlin,
The Kaiser he will say,
'Hoch, hoch! Mein Gott,
What a bloody fine lot
Are the ragtime infantry!'

(iii)

SURE, a little bit of shrapnel fell from out the sky one day,
And it nestled in my shoulder in a quaint and loving way,
And when the doctor saw it, it looked so sweet and fair,
He said, 'Suppose we leave it for it looks so peaceful there'.
Then he painted it with iodine to keep the germs away,
It's the only way to treat it, no matter what they say.
But early the next morning he changed his fickle mind,
And he marked me down for duty and he sent me up the line.

Fred Karno] popular comedian

(iv)

I HAVE no pain, dear mother, now,
But oh! I am so dry.
Connect me to a brewery
And leave me there to die.

(v)

WASH me in the water
That you washed your dirty daughter,
And I shall be whiter
Than the whitewash on the wall.
Whiter
Than the whitewash on the wall,
Whiter
Than the whitewash on the wall.
Oh, wash me in the water
That you washed your dirty daughter,
And I shall be whiter
Than the whitewash on the wall.

(vi)

THE bells of hell go ting-a-ling-a-ling
For you but not for me:
And the little devils how they sing-a-ling-a-ling
For you but not for me.
O Death, where is thy sting-a-ling-a-ling,
O Grave, thy victor-ee?
The bells of hell go ting-a-ling-a-ling,
For you but not for me.

SAMUEL HOFFENSTEIN

1890–1947

292

Progress

THEY'LL soon be flying to Mars, I hear—
But how do you open a bottle of beer?

A flash will take you from Nome to New York—
But how the hell do you pull a cork?

They'll rocketeer you to Hibernia—
But open a window and get a hernia.

They've stripped space from the widow'd blue—
But where is the lace that fits a shoe?

Where is the key that fits a lock?
Where is the garter that holds a sock?

They'll hop to the moon and skip to the stars,
But what'll stay put are the lids on jars.

The mighty telescope looks far,
But finds no place to park a car.

The world crackles with cosmic minds
Tangled up in Venetian blinds.

One day they'll resurrect the dead,
Who'll die again of colds in the head.

293 *I'm Fond of Doctors*

I'M fond of doctors and drivers of hacks
Whose names are Morris and Barney and Max;
I'm fond of waiters in places I know
Whose names are Louis and Mike and Joe;
They take my mind off taxes and love—
A very good taking the mind off of.

294 from *Love-songs, at Once Tender and Informative*

YOUR little hands,
Your little feet,
Your little mouth—
Oh, God, how sweet!

Your little nose,
Your little ears,
Your eyes, that shed
Such little tears!

Your little voice,
So soft and kind;
Your little soul,
Your little mind!

295 from *Songs of Fairly Utter Despair*

NOW, alas, it is too late
To buy Manhattan real estate,
But when my father came to town,
He could have bought for fifty down,
And I should not be where I am:
Yet does my father give-a-damn,
Or ever say, 'I'm sorry, boy,'
Or looking at me, murmur, 'Oy?'
He does not grieve for what I've missed,
And yet I'm called an Anarchist!

296 from *Poems in Praise of Practically Nothing*

(i)

YOU buy some flowers for your table;
You tend them tenderly as you're able;
You fetch them water from hither and thither—
What thanks do you get for it all? They wither.

(ii)

You buy yourself a new suit of clothes;
The care you give it, God only knows;
The material, of course, is the very *best* yet;
You get it pressed and pressed and *pressed* yet;
You keep it free from specks *so* tiny—
What thanks do you get? The pants get shiny.

(iii)

You hire a cook, but she can't cook yet;
You teach her by candle, bell, and book yet;
You show her, as if she were in her cradle,
Today, the soup, tomorrow, a ladle.
Well, she doesn't learn, so although you need her,
You decide that somebody else should feed her:—

But you're kind by birth; you hate to fire her;
To tell a woman you don't require her—
So you wait and wait, and before you do it,
What thanks do you get? She beats you to it!

ANONYMOUS

297

The Pig

It was an evening in November,
As I very well remember,
I was strolling down the street in drunken pride,
But my knees were all a-flutter,
And I landed in the gutter
And a pig came up and lay down by my side.

Yes, I lay there in the gutter
Thinking thoughts I could not utter,
When a colleen passing by did softly say
'You can tell a man who boozes
By the company he chooses'—
And the pig got up and slowly walked away.

COLE PORTER

1891–1964

298

I'm a Gigolo

I should like you all to know,
I'm a famous gigolo.
And of lavender, my nature's got just a dash in it.
As I'm slightly undersexed,
You will always find me next
To some dowager who's wealthy rather than passionate.

Go to one of those night club places
And you'll find me stretching my braces
Pushing ladies with lifted faces 'round the floor.
But I must confess to you
There are moments when I'm blue.
And I ask myself whatever I do it for.

I'm a flower that blooms in the winter.
Sinking deeper and deeper in 'snow.'
I'm a baby who has
No mother but jazz,
I'm a gigolo.
Ev'ry morning, when labor is over,
To my sweet-scented lodgings I go,
Take the glass from the shelf
And look at myself,
I'm a gigolo.
I get stocks and bonds
From faded blondes
Ev'ry twenty-fifth of December.
Still I'm just a pet
That men forget
And only tailors remember.
Yet when I see the way all the ladies
Treat their husbands who put up the dough.
You cannot think me odd
If then I thank God
I'm a gigolo.

299 *Brush Up your Shakespeare*

THE girls today in society
Go for classical poetry
So to win their hearts one must quote with ease
Aeschylus and Euripides.
One must know Homer and, b'lieve me, bo,
Sophocles, also Sappho-ho.
Unless you know Shelley and Keats and Pope,
Dainty debbies will call you a dope.
But the poet of them all
Who will start 'em simply ravin'
Is the poet people call
'The bard of Stratford-on-Avon.'

Brush up your Shakespeare,
Start quoting him now,
Brush up your Shakespeare
And the women you will wow.
Just declaim a few lines from 'Othella'
And they'll think you're a helluva fella,
If your blonde won't respond when you flatter 'er
Tell her what Tony told Cleopaterer,
If she fights when her clothes you are mussing,
What are clothes? 'Much Ado About Nussing.'
Brush up your Shakespeare
And they'll all kowtow.

Brush up your Shakespeare,
Start quoting him now,
Brush up your Shakespeare
And the women you will wow.
With the wife of the British embessida
Try a crack out of 'Troilus and Cressida,'
If she says she won't buy it or tike it
Make her tike it, what's more, 'As You Like It.'
If she says your behavior is heinous
Kick her right in the 'Coriolanus,'
Brush up your Shakespeare
And they'll all kowtow.

Brush up your Shakespeare,
Start quoting him now,
Brush up your Shakespeare
And the women you will wow.
If you can't be a ham and do 'Hamlet'
They will not give a damn or a damnlet,
Just recite an occasional sonnet
And your lap'll have 'Honey' upon it,
When your baby is pleading for pleasure
Let her sample your 'Measure for Measure,'
Brush up your Shakespeare
And they'll all kowtow.

Brush up your Shakespeare,
Start quoting him now,
Brush up your Shakespeare
And the women you will wow.
Better mention 'The Merchant of Venice'
When her sweet pound o' flesh you would menace,

If her virtue, at first, she defends—well,
Just remind her that 'All's Well That Ends Well,'
And if still she won't give you a bonus
You know what Venus got from Adonis!
Brush up your Shakespeare
And they'll all kowtow.

Brush up your Shakespeare
Start quoting him now,
Brush up your Shakespeare
And the women you will wow.
If your goil is a Washington Heights dream
Treat the kid to 'A Midsummer Night's Dream,'
If she then wants an all-by-herself night
Let her rest ev'ry 'leventh or 'Twelfth Night,'
If because of your heat she gets huffy
Simply play on and 'Lay on, Macduffy!'
Brush up your Shakespeare
And they'll all kowtow.

Brush up your Shakespeare,
Start quoting him now,
Brush up your Shakespeare
And the women you will wow.
So tonight just recite to your matey
'Kiss me, Kate, Kiss me, Kate, Kiss me, Katey,'
Brush up your Shakespeare
And they'll all kowtow.

MORRIS BISHOP

1893–1973

300

My Friend the Cuckold

I KNEW a cuckold once. I grieve for him.
  When his wife's sin was told him by some tattler,
The news made all existence bleak and grim
  And wrecked his home, which was, in fact, the Statler.

'A horsewhip!' he kept shouting. 'Yes, I swear
I'll horsewhip that seducer so abhorred!'
He could not buy a horsewhip anywhere,
Not from Sears Roebuck nor Montgomery Ward.

For farm and stock whips, drovers' whips, and quirts
Alone are catalogued. 'It is my ruin!'
He cried. 'The horsewhip heals our honor's hurts.
Who ever heard of quirting a Don Juan?'

He sought relief in drink, which made him ill.
'*Je suis cocu!*' he would complain, demanding
One's sympathy. 'I use the French, for still
In France the cuckold has a certain standing.

'But here the general public does not know—'
And somewhat horribly he gasped and chuckled—
'Even the right pronunciation. Oh,
It is not gay, my friend, to be a cuckold!'

301 *The Adventures of Id*

'Oh you kid!'
Shouted Id.
'Take it easy, amigo,'
Muttered Ego.

'Babe, we'll take the lid
Off the town!' said Id.
'What will folks say in Oswego?'
Said Ego.

'Babe, let's get rid
Of this killjoy,' said Id.
'If I go, will she go?'
Said Ego.

Vast and calm and politic
Super-Ego swung his stick.
Super-Ego turned his head:
'You boys better get back to bed.'
'Well, I never did!'
Said Id.

J. B. MORTON ('BEACHCOMBER')

1893–1979

302

On Sir Henry Ferrett, MP

'SEEING is believing,'
I've often heard you say.
My dear Sir Henry Ferrett,
I see you every day.

303

To Hilda Dancing

WHEN Hilda does the Highland reel,
She steals my heart away;
No dancing bear is so genteel,
Or half so *dégagée*.

304

Spring in London

LET poets praise the blossom of wild Spring,
These cleaner skies, this magic-laden air;
I mark the season by a greater thing—
Viscountess Trasche is back in Berkeley Square.

DOROTHY PARKER

1893–1967

305

One Perfect Rose

A SINGLE flow'r he sent me, since we met.
All tenderly his messenger he chose;
Deep-hearted, pure, with scented dew still wet—
One perfect rose.

I knew the language of the floweret;
  'My fragile leaves,' it said, 'his heart enclose.'
Love long has taken for his amulet
  One perfect rose.

Why is it no one ever sent me yet
  One perfect limousine, do you suppose?
Ah no, it's always just my luck to get
  One perfect rose.

306

Comment

OH, life is a glorious cycle of song,
A medley of extemporanea;
And love is a thing that can never go wrong;
And I am Marie of Roumania.

GERALD BULLETT

1893–1958

307

To Archbishop Lang

(Written at the time of the Abdication, 1936)

MY Lord Archbishop, what a scold you are,
And when your man is down how bold you are,
Of Christian charity how scant you are—
How Lang, O Lord, how full of cant you are!

PHILIP HESELTINE

1894–1930

308

Picture-Palaces

THE young things who frequent picture-palaces
Have no time for psychoanalysis,
   And though Dr Freud
   Is distinctly annoyed
They cling to their long-standing fallacies.

E. E. CUMMINGS

1894–1962

309

the Noster

the Noster was a ship of swank
(as gallant as they come)
until she hit a mine and sank
just off the coast of Sum

precisely where a craft of cost
the Ergo perished later
all hands (you may recall) being lost
including captain Pater

310

may i feel said he

may i feel said he
(i'll squeal said she
just once said he)
it's fun said she

(may i touch said he
how much said she
a lot said he)
why not said she

(let's go said he
not too far said she
what's too far said he
where you are said she)

may i stay said he
(which way said she
like this said he
if you kiss said she

may i move said he
is it love said she)
if you're willing said he
(but you're killing said she

but it's life said he
but your wife said she
now said he)
ow said she

(tiptop said he
don't stop said she
oh no said he)
go slow said she

(cccome? said he
ummm said she)
you're divine! said he
(you are Mine said she)

311 *mr u*

mr u will not be missed
who as an anthologist
sold the many on the few
not excluding mr u

ALDOUS HUXLEY

1894–1963

312

Second Philosopher's Song

IF, O my Lesbia, I should commit,
Not fornication, dear, but suicide,
My Thames-blown body (Pliny vouches it)
Would drift face upwards on the oily tide
With the other garbage, till it putrefied.

But you, if all your lovers' frozen hearts
Conspired to send you, desperate, to drown—
Your maiden modesty would float face down,
And men would weep upon your hinder parts.

'Tis the Lord's doing. Marvellous is the plan
By which this best of worlds is wisely planned.
One law He made for woman, one for man:
We bow the head and do not understand.

313

Fifth Philosopher's Song

A MILLION million spermatozoa,
   All of them alive:
Out of their cataclysm but one poor Noah
   Dare hope to survive.

And among that billion minus one
   Might have chanced to be
Shakespeare, another Newton, a new Donne—
   But the One was Me.

Shame to have ousted your betters thus,
   Taking ark while the others remained outside!
Better for all of us, froward Homunculus,
   If you'd quietly died!

314 from *Antic Hay*

CHRISTLIKE is my behaviour,
Like every good believer,
I imitate the Saviour,
And cultivate a beaver.

ROBERT GRAVES

1895–1985

315 *Beauty in Trouble*

BEAUTY in trouble flees to the good angel
On whom she can rely
To pay her cab-fare, run a steaming bath,
Poultice her bruised eye;

Will not at first, whether for shame or caution,
Her difficulty disclose;
Until he draws a cheque book from his plumage,
Asking how much she owes.

(Breakfast in bed: coffee and marmalade,
Toast, eggs, orange-juice,
After a long, sound sleep—the first since when?—
And no word of abuse.)

Loves him less only than her saint-like mother,
Promises to repay
His loans and most seraphic thoughtfulness
A million-fold one day.

Beauty grows plump, renews her broken courage
And, borrowing ink and pen,
Writes a news-letter to the evil angel
(Her first gay act since when?):

The fiend who beats, betrays and sponges on her,
  Persuades her white is black,
Flaunts vespertilian wing and cloven hoof;
  And soon will fetch her back.

Virtue, good angel, is its own reward:
  Your guineas were well spent.
But would you to the marriage of true minds
  Admit impediment?

316

Traveller's Curse After Misdirection

(Translation from the Welsh)

MAY they stumble, stage by stage
On an endless pilgrimage,
Dawn and dusk, mile after mile,
At each and every step, a stile;
At each and every step withal
May they catch their feet and fall;
At each and every fall they take
May a bone within them break;
And may the bone that breaks within
Not be, for variation's sake,
Now rib, now thigh, now arm, now shin,
But always, without fail, THE NECK.

317

A Grotesque

MY Chinese uncle, gouty, deaf, half-blinded,
And more than a trifle absent-minded,
Astonished all St James's Square one day
By giving long and unexceptionably exact directions
To a little coolie girl, who'd lost her way.

318

The Weather of Olympus

ZEUS was once overheard to shout at Hera:
  'You hate it, do you? Well, I hate it worse—
East wind in May, sirocco all the Summer.
  Hell take this whole impossible Universe!'

A scholiast explains his warm rejoinder,
  Which sounds too man-like for Olympic use,
By noting that the snake-tailed Chthonian winds
  Were answerable to Fate alone, not Zeus.

319 *1805*

At Viscount Nelson's lavish funeral,
  While the mob milled and yelled about St Paul's,
A General chatted with an Admiral:

'One of your Colleagues, Sir, remarked today
  That Nelson's *exit*, though to be lamented,
Falls not inopportunely, in its way.'

'He was a thorn in our flesh,' came the reply—
  'The most bird-witted, unaccountable,
Odd little runt that ever I did spy.

'One arm, one peeper, vain as Pretty Poll,
  A meddler, too, in foreign politics
And gave his heart in pawn to a plain moll.

'He would dare lecture us Sea Lords, and then
  Would treat his ratings as though men of honour
And play at leap-frog with his midshipmen!

'We tried to box him down, but up he popped,
  And when he'd banged Napoleon at the Nile
Became too much the hero to be dropped.

'You've heard that Copenhagen "blind eye" story?
  We'd tied him to Nurse Parker's apron-strings—
By G——d, he snipped them through and snatched the glory!'

'Yet,' cried the General, 'six-and-twenty sail
  Captured or sunk by him off Trafalgar—
That writes a handsome *finis* to the tale.'

'Handsome enough. The seas are England's now.
  That fellow's foibles need no longer plague us.
He died most creditably, I'll allow.'

'And, Sir, the secret of his victories?'
'By his unServicelike, familiar ways, Sir,
He made the whole Fleet love him, damn his eyes!'

320 *Epitaph on an Unfortunate Artist*

HE found a formula for drawing comic rabits:
The formula for drawing comic rabbits paid,
So in the end he could not change the tragic habits
This formula for drawing comic rabbits made.

321 *Twins*

SIAMESE twins: one, maddened by
The other's moral bigotry,
Resolved at length to misbehave
And drink them both into the grave.

EDMUND WILSON

1895–1972

322 *Disloyal Lines to an Alumnus*

Who wrote poetry about coming back to college 'like a man to his mother returning' and feeling 'the keen swift faith that God is good,' and who later complained in the *Alumni Weekly* that books by alumni authors were not being sufficiently praised by the reviews in that periodical.

I, TOO, have faked the glamor of gray towers,
I, too, have sung the ease of sultry hours—
Deep woods, sweet lanes, wide playing fields, smooth ponds
—Where clean boys train to sell their country's bonds.
Ah, high delights untasted by outsiders!
The Graduate College with its dreaming spiders!
May windows to the summer drunks flung wide!
The ivied peal of bells at eventide!
The drone of doves in immemorial trees,
The bumble of innumerable bees!—

And Beauty, Beauty, oozing everywhere
Like maple-sap from maples! Dreaming there,
I have sometimes stepped in Beauty on the street
And slipped, sustaining bruises blue but sweet,
And felt the keen swift faith, I will assert,
That God is fairly good to Struthers Burt!

—For God and Struthers Burt are gentle folks:
They differ from Jack Dempsey and Joe Doaks.
God is a big beneficent trustee,
Who asks well-bred professors in to tea;
Has swans and swimming-pools about the grounds;
Collects old clocks, and sometimes rides to hounds.
God was a club or two ahead of Burt,
But not enough to make him cold or curt—
They both believe in college comradeship,
Old college ways, the slow delicious drip
Of cool damp verse; and also, I suppose,
The keen and peevish tang of high-pitched prose.
Burt sometimes goes to stay with God for weeks
And utters fierce shrill Philadelphian squeaks.

323 from *Easy Exercises in the Use of Difficult Words*

(i)

Nursery Vignette

THE bubbled baby gave an abrupt burp,
Her tiny face contorted in an irpe
(The *i* pronounced, perhaps, like *beard* not bird).
Ben Jonson only used this pleasant little word.

(ii)

Scène de boudoir

SAID Philip Sidney, buttoning his jerkin,
'Allow me, darling: you have dropped your merkin.'

(iii)

Lakeside

AN old cob swan his cygnets thus addressed:
'Stray not too far from the parental nest.
Remember you can never be as spry as is
Yon falcon with her eyrie full of eyases!'

324

Enemies of Promise

CYRIL CONNOLLY
Behaves rather fonnily:
Whether folks are at peace or fighting,
He complains that it keeps him from writing.

325

Miniature Dialogue

SAID Mario Praz to Mario Pei,
'Che cosa noiosa the Great White Way!'
'But full of delightful polyglots!'
Said Mario Pei to Mario Praz.

LORENZ HART

1895–1943

326

I Wish I Were in Love Again

THE sleepless nights,
The daily fights,
The quick toboggan when you reach the heights—
I miss the kisses and I miss the bites.
I wish I were in love again!
The broken dates,
The endless waits,
The lovely loving and the hateful hates,
The conversation with the flying plates—
I wish I were in love again!
No more pain,
No more strain,
Now I'm sane, but . . .
I would rather be gaga!
The pulled-out fur of cat and cur,
The fine mismating of a him and her—
I've learned my lesson, but I
Wish I were in love again.

Mario Praz] aesthete and literary historian *Mario Pei*] linguist and philologist

The furtive sigh,
The blackened eye,
The words 'I'll love you till the day I die',
The self-deception that believes the lie—
I wish I were in love again.
When love congeals
It soon reveals
The faint aroma of performing seals,
The double-crossing of a pair of heels.
I wish I were in love again!
No more care.
No despair.
I'm all there now,
But I'd rather be punch-drunk!
Believe me, sir,
I much prefer
The classic battle of a him and her.
I don't like quiet and I
Wish I were in love again!

L. A. G. STRONG

1896–1958

327 *The Brewer's Man*

HAVE I a wife? Bedam I have!
But we was badly mated.
I hit her a great clout one night,
And now we're separated.
And mornings going to me work
I meets her on the quay:
'Good mornin' to ye, ma'am!' says I,
'To hell with ye!' says she.

IRA GERSHWIN

1896–1983

328 *The Babbitt and the Bromide*

1

A BABBITT met a Bromide on the avenue one day.
They held a conversation in their own peculiar way.
They both were solid citizens—they both had been around.
And as they spoke you clearly saw their feet were on the ground:

Hello! How are you?
Howza folks? What's new?
I'm great! That's good!
Ha! Ha! Knock wood!
Well! Well! What say?
Howya been? Nice day!
How's tricks? What's new?
That's fine! How are you?
Nice weather we are having but it gives me such a pain:
I've taken my umbrella, so of course it doesn't rain.
Heigh ho! That's life!
What's new? Howza wife?
Gotta run! Oh, my!
Ta! Ta! Olive oil! Good bye!

2

Ten years went quickly by for both these sub-sti-an-tial men,
Then history records one day they chanced to meet again.
That they had both developed in ten years there was no doubt,
And so of course they had an awful lot to talk about:

Hello! How are you? &c.
I'm sure I know your face, but I just can't recall your name;
Well, how've you been, old boy, you're looking just about the same.
Heigh ho! That's life! &c.

3

Before they met again some twenty years they had to wait.
This time it happened up above, inside St Peter's gate.
A harp each one was carrying and both were wearing wings,
And this is what they sang as they kept strumming on the strings:

Hello! How are you? &c.
You've grown a little stouter since I saw you last, I think.
Come up and see me sometime and we'll have a little drink.
Heigh ho! That's life! &c.

SAGITTARIUS
(Olga Katzin)

1896–1987

329

Stalin Moy Golubchik

(Overheard at the Carlton Club)

OH! Stalin is my darling, my darling, my darling,
Stalin is my darling, the old *molodyetz.*

I much rejoice to have a choice of Russian epithets,
And spread the fame of Stalin's name as a true *molodyetz.*

A real good sport, he holds the fort and laughs at Hitler's threats,
He pulls his weight for Church and State, the old *molodyetz.*

Though they are Red, much may be said for Fighting Soviets—
One Communist I can't resist, the old *molodyetz.*

The *status quo* of long ago I drop without regrets,
Allied we fight for Red and Right with the old *molodyetz.*

We will supply this great Ally forgetting loans or debts,
Give lend and lease (until the peace) to the old *molodyetz.*

Stalin moy golubchik, golubchik, golubchik,
Stalin, moy golubchik, the old *molodyetz.*

molodyetz] good fellow

330

The Passionate Profiteer to his Love

(After Christopher Marlowe)

COME feed with me and be my love,
And pleasures of the table prove,
Where *Prunier* and *The Ivy* yield
Choice dainties of the stream and field.

At *Claridge* thou shalt duckling eat,
Sip vintages both dry and sweet,
And thou shalt squeeze between thy lips
Asparagus with buttered tips.

On caviare my love shall graze,
And plump on salmon mayonnaise,
And browse at *Scott's* beside thy swain
On lobster Newburg with champagne.

Between hors d'œuvres and canapés
I'll feast thee on *poularde soufflé*
And every day within thy reach
Pile melon, nectarine and peach.

Come share at the *Savoy* with me
The menu of austerity;
If in these pastures thou wouldst rove
Then feed with me and be my love.

DAVID McCORD

1897–

331

Epitaph on a Waiter

BY and by
God caught his eye.

332

When I Was Christened

WHEN I was christened
they held me up
and poured some water
out of a cup.

The trouble was
it fell on me,
and I and water
don't agree.

A lot of christeners
stood and listened:
I let them know
that I was christened.

C. S. LEWIS

1898–1963

333

Ballade of Dead Gentlemen

WHERE, in what bubbly land, below
  What rosy horizon dwells to-day
That worthy man Monsieur Cliquot
  Whose widow has made the world so gay?
  Where now is Mr Tanqueray?
Where might the King of Sheba be
  (Whose wife stopped dreadfully long away)?
Mais où sont messieurs les maris?

Say where did Mr Beeton go
  With rubicund nose and whiskers grey
To dream of dumplings long ago,
  Of syllabubs, soups, and *entremets*?
  In what dim isle did Twankey lay
His aching head? What murmuring sea
  Lulls him after the life-long fray?
Mais où sont messieurs les maris?

How Mr Grundy's cheeks may glow
  By a bathing-pool where lovelies play,
I guess, but shall I ever know?
  Where—if it comes to that, *who*, pray—
  Is Mr Masham? Sévigné
And Mr Siddons and Zebedee
  And Gamp and Hemans, where are they?
Mais où sont messieurs les maris?

Princesses all, beneath your sway
  In this grave world they bowed the knee;
Libertine airs in Elysium say
  Mais où sont messieurs les maris?

334

An Epitaph

ERECTED by her sorrowing brothers
In memory of Martha Clay.
Here lies one who lived for others;
Now she has peace. And so have they.

W. C. SELLAR (1898–1951) and R. J. YEATMAN (1898–1968)

335

Old-Saxon Fragment

SYNG a song of Saxons
In the Wapentake of Rye
Four and twenty eaoldormen
Too eaold to die . . . .

336 *Beoleopard*, or *The Witan's Whail*

WHAN Cnut Cyng the Witan wold enfeoff
Of infangthief and outfangthief
Wonderlich were they enwraged
And wordwar waged
Sware Cnut great scot and lot
Swingë wold ich this illbegotten lot.

Wroth was Cnut and wrothword spake.
Well wold he win at wopantake.
Fain wold he brakë frith and crackë heads
And than they shold worshippe his redes.

Swinged Cnut Cyng with swung sword
Howled Witanë hellë but hearkened his word
Murie sang Cnut Cyng
Outfangthief is Damgudthyng.

NOËL COWARD

1899–1973

337 *The Stately Homes of England*

LORD ELDERLEY, Lord Borrowmere, Lord Sickert and
 Lord Camp,
With every virtue, every grace,
Ah! what avails the sceptred race.
Here you see the four of us,
And there are so many more of us,
Eldest sons that must succeed.
We know how Caesar conquered Gaul
And how to whack a cricket ball,
Apart from this our education
Lacks co-ordination.
Tho' we're young and tentative
And rather rip-representative
Scions of a noble breed,
We are the products of those homes serene and stately
Which only lately
Seem to have run to seed!

The Stately Homes of England
How beautiful they stand,
To prove the upper classes
Have still the upper hand;
Tho' the fact that they have to be rebuilt
And frequently mortgaged to the hilt
Is inclined to take the gilt
Off the gingerbread,
And certainly damps the fun
Of the eldest son.
But still we won't be beaten,
We'll scrimp and screw and save,
The playing-fields of Eton
Have made us frightfully brave,
And tho' if the Van Dycks have to go
And we pawn the Bechstein Grand,
We'll stand by the Stately Homes of England.

Here you see the pick of us,
You may be heartily sick of us
Still with sense we're all imbued.
We waste no time on vain regrets
And when we're forced to pay our debts
We're always able to dispose of
Rows and rows and rows of
Gainsboroughs and Lawrences,
Some sporting prints of Aunt Florence's,
Some of which are rather rude.
Altho' we sometimes flaunt our family conventions,
Our good intentions
Mustn't be misconstrued.

The Stately Homes of England
We proudly represent,
We only keep them up for
Americans to rent.
Tho' the pipes that supply the bathroom burst
And the lavatory makes you fear the worst,
It was used by Charles the First
Quite informally,
And later by George the Fourth
On a journey North.

The State Apartments keep their
Historical renown,
It's wiser not to sleep there
In case they tumble down;
But still if they ever catch on fire
Which, with any luck, they might,
We'll fight for the Stately Homes of England.

The Stately Homes of England,
Tho' rather in the lurch,
Provide a lot of chances
For psychical research.
There's the ghost of a crazy younger son
Who murdered in Thirteen Fifty One
An extremely rowdy nun
Who resented it,
And people who come to call
Meet her in the hall.
The baby in the guest wing
Who crouches by the grate,
Was walled up in the west wing
In Fourteen Twenty Eight.
If anyone spots the Queen of Scots
In a hand-embroidered shroud,
We're proud of the Stately Homes of England.

338 *Irish Song*

WHEN first I was courtin' sweet Rosie O'Grady,
Sweet Rosie O'Grady she whispered to me,
'Sure you shouldn't be after seducin' a lady
Before she's had time to sit down to her tea.'

With a Heigho—Top-o-the-morning—Begorrah and Fiddlededee.

Her cheeks were so soft and her eyes were so trustin',
She tossed her bright curls at the dusk of the day,
She said to me, 'Darlin', your breath is disgustin',
Which wasn't at all what I hoped she would say.

With a Heigho, maybe Begorrah, and possibly Fiddlededee.

Our honeymoon started so blithely and gaily
But dreams I was dreaming were suddenly wrecked
For she broke my front tooth with her father's shillelagh
Which wasn't what I had been led to expect.

With a Heigho, maybe Begorrah, and certainly Fiddlededee.

JOSEPH MONCURE MARCH

1899–1977

339 from *The Wild Party*

QUEENIE was a blonde, and her age stood still,
And she danced twice a day in vaudeville.
Grey eyes.
Lips like coals aglow.
Her face was a tinted mask of snow.
What hips—
What shoulders—
What a back she had!
Her legs were built to drive men mad.
And she did.
She would skid.
But sooner or later they bored her:
Sixteen a year was her order.

They might be blackguards;
They might be curs;
They might be actors; sports; chauffeurs—
She never inquired
Of the men she desired
About their social status, or wealth:
She was only concerned about their health.
True:
She knew:
There was little she hadn't been through.

And she liked her lovers violent, and vicious:
Queenie was sexually ambitious.
So:
Now you know.
A fascinating woman, as they go.

She lived at present with a man named Burrs
Whose act came on just after hers.
A clown
Of renown:
Three-sheeted all over town.
He was comical as sin;
Comical as hell;
A gesture—a grin,
And the house would yell,
Uproarious:
He was glorious!

So from the front. People in the wings
Saw him and thought of other things
Coldly—
Most coldly:
Many would say them boldly,
Adding in language without much lace
They'd like to break his god-damned face.
Ask why?
They might be stuck:
They would like to, just for luck.

But these were men, for the greater part.
A woman would offer him up her heart
Throbbing,
On a platter:
He could bite it, and it wouldn't matter.
As long as he kissed, and held her tight,
And gave her a fairly hectic night.
Which he could,
And would.
A man these women understood!

Oh, yes—Burrs was a charming fellow:
Brutal with women, and proportionately yellow.
Once he had been forced into a marriage.
Unlucky girl!
She had a miscarriage

Two days later. Possibly due
To the fact that Burrs beat her with the heel of a shoe
Till her lips went blue.
For a week, her brother had great fun
Looking for Burrs with a snub-nosed gun:
At the end of which time, she began to recover;
And Burrs having vanished, the thing blew over.
Just a sample
For example:
One is probably ample.

*

Studio;
Bedroom;
Bath;
Kitchenette:
Furnished like a third act passion set:
Oriental;
Sentimental;
They owed two months on the rental.
Pink cushions,
Blue cushions: overlaid
With silk: with lace: with gold brocade.
These lay propped up on a double bed
That was covered with a Far East tapestry spread.

Chinese dragons with writhing backs:
Photographs caught to the wall with tacks:
Their friends in the profession,
Celebrities for the impression—
('So's your old man—Isidore.'
'Faithfully—Ethel Barrymore.')

On a Chinese lacquer tray there stood a
Gong with tassels, and a brass Buddha.
Brass candlesticks.
Orange candles.
An Art vase with broken handles,
Out of which came an upthrusting
Of cherry blossoms that needed dusting.

Books?
Books?
My god! You don't understand.
They were far too busy living first-hand
For books.
Books!

True,
On the table there lay a few
Tattered copies of a magazine,
Confessional;
Professional;
That talked of their friends on the stage and screen.

A Victrola with records
Just went to show
Queenie's Art on the man two floors below.
Being a person of little guile,
He had lent them to her, for just awhile.
Believe it or not—
All this for a smile!

A grand piano stood in the corner
With the air of a coffin waiting for a mourner.

The bath was a horrible give-away.
The floor was dirty:
The towels were grey.
Cups, saucers,
Knives, plates,
Bottles, glasses
In various states
Of vileness, fought for precarious space in
The jumbled world beneath the basin.
The basin top was the temporary home
Of a corkscrew, scissors,
And a brush and comb.
In the basin bowel
Was a Pullman towel
Vividly wrought with red streaks
From Queenie's perfect lips and cheeks.
Behind one faucet, in a stain of rust,
Spattered with talcum powder and dust,
A razor blade had lived for weeks.
Beside it was stuck a cigarette stub.
And the tub?
Oh—never mind the tub!
On the door-knob there hung a pair
Of limp stockings, and a brassière
Too soiled to wear.

Of the bedroom,
Nothing much to be said.
It had a bureau:
A double bed
With one pillow, and white spread.
Their trunks: boxes.
A chair.
The walls were white and bare.
Only occasional guests slept there:
Queenie and Burrs, preferring air,
Slept with the Chinese dragons instead.

ANONYMOUS

340

A Little Lamb

MARY had a little lamb,
She ate it with mint sauce,
And everywhere that Mary went
The lamb went too, of course.

ANONYMOUS

341

Advertising Rhymes

(i)

THEY come as a boon and a blessing to men,
The Pickwick, the Owl, and the Waverley pen.

(ii)

Force

JIM DUMPS was a most unfriendly man,
Who lived his life on the hermit plan.
In his gloomy way he'd gone through life
And made the most of woe and strife,
Till Force one day was served to him.
Since then they've called him Sunny Jim.

(iii)

THERE was a little man,
And he felt a little glum.
He thought that a Guinness was due, due, due.
So he went to 'The Plough' . . .
And he's feeling better now.
For a Guinness is good for you, you, you.

(iv)

Burma-Shave Roadside Signs

WITHIN THIS VALE
OF TOIL
AND SIN
YOUR HEAD GROWS BALD
BUT NOT YOUR CHIN—USE
BURMA-SHAVE

NO LADY LIKES
TO DANCE
OR DINE
ACCOMPANIED BY
A PORCUPINE
BURMA-SHAVE

BE A MODERN
PAUL REVERE
SPREAD THE NEWS
FROM EAR
TO EAR
BURMA-SHAVE

THE BURMA GIRLS
IN MANDALAY
DUNK BEARDED LOVERS
IN THE BAY
WHO DON'T USE
BURMA-SHAVE

THE QUEEN
OF HEARTS
NOW LOVES THE KNAVE
THE KING
RAN OUT OF
BURMA-SHAVE

MY JOB IS
KEEPING FACES CLEAN
AND NOBODY KNOWS
DE STUBBLE
I'VE SEEN
BURMA-SHAVE

HARDLY A DRIVER
IS NOW ALIVE
WHO PASSED
ON HILLS
AT 75
BURMA-SHAVE

BROTHER SPEEDERS
LET'S
REHEARSE
ALL TOGETHER
'GOOD MORNING, NURSE!'
BURMA-SHAVE

A GUY
WHO WANTS
TO MIDDLE-AISLE IT
MUST NEVER SCRATCH
HIS LITTLE VIOLET
BURMA-SHAVE

IT GAVE
MCDONALD
THAT NEEDED CHARM
HELLO HOLLYWOOD
GOOD-BY FARM
BURMA-SHAVE

OGDEN NASH

1902–1971

342

The Emmet

THE emmet is an ant (archaic),
The ant is just a pest (prosaic).
The modern ant, when trod upon,
Exclaims 'I'll be a son-of-a-gun!'
Not so its ancestor, the emmet,
Which perished crying 'Zounds!' or 'Demmit!'

343

The Fly

GOD in His wisdom made the fly
And then forgot to tell us why.

344

Introspective Reflection

I WOULD live all my life in nonchalance and insouciance
Were it not for making a living, which is rather a nouciance.

345

Curl Up and Diet

SOME ladies smoke too much and some ladies drink too much and some ladies pray too much,
But all ladies think that they weigh too much.
They may be as slender as a sylph or a dryad,
But just let them get on the scales and they embark on a doleful jeremiad:
No matter how low the figure the needle happens to touch,
They always claim it is at least five pounds too much;
To the world she may appear slinky and feline,
But she inspects herself in the mirror and cries, Oh, I look like a sea lion.
Yes, she tells you she is growing into the shape of a sea cow or manatee,
And if you say No, my dear, she says you are just lying to make her feel better, and if you say Yes, my dear, you injure her vanity.
Once upon a time there was a girl more beautiful and witty and charming than tongue can tell,
And she is now a dangerous raving maniac in a padded cell,

And the first indication her friends and relatives had that she was mentally overwrought
Was one day when she said, I weigh a hundred and twenty-seven, which is exactly what I ought.
Oh, often I am haunted
By the thought that somebody might someday discover a diet that would let ladies reduce just as much as they wanted,
Because I wonder if there is a woman in the world strong-minded enough to shed ten pounds or twenty,
And say There now, that's plenty;
And I fear me one ten-pound loss would only arouse the craving for another,
So it wouldn't do any good for ladies to get their ambition and look like somebody's fourteen-year-old brother,
Because, having accomplished this with ease,
They would next want to look like somebody's fourteen-year-old brother in the final stages of some obscure disease,
And the more success you have the more you want to get of it,
So then their goal would be to look like somebody's fourteen-year-old brother's ghost, or rather not the ghost itself, which is fairly solid, but a silhouette of it,
So I think it is very nice for ladies to be lithe and lissome.
But not so much so that you cut yourself if you happen to embrace or kissome.

346 *Samson Agonistes*

I TEST my bath before I sit.
And I'm always moved to wonderment
That what chills the finger not a bit
Is so frigid upon the fundament.

347 *The Private Dining-Room*

MISS RAFFERTY wore taffeta,
Miss Cavendish wore lavender.
We ate pickerel and mackerel
And other lavish provender.
Miss Cavendish was Lalage,
Miss Rafferty was Barbara.
We gobbled pickled mackerel
And broke the candelabara,

Miss Cavendish in lavender,
In taffeta, Miss Rafferty,
The girls in taffeta lavender,
And we, of course, in mufti.

Miss Rafferty wore taffeta,
The taffeta was lavender,
Was lavend, lavender, lavenderest,
As the wine improved the provender.
Miss Cavendish wore lavender,
The lavender was taffeta.
We boggled mackled pickerel,
And bumpers did we quaffeta.
And Lalage wore lavender,
And lavender wore Barbara,
Rafferta taffeta Cavender lavender
Barbara abracadabra.

Miss Rafferty in taffeta
Grew definitely raffisher.
Miss Cavendish in lavender
Grew less and less stand-offisher.
With Lalage and Barbara
We grew a little pickereled,
We ordered Mumm and Roederer
Because the bubbles tickereled.
But lavender and taffeta
Were gone when we were soberer.
I haven't thought for thirty years
Of Lalage and Barbara.

348

Grandpa Is Ashamed

A CHILD need not be very clever
To learn that 'Later, dear' means 'Never.'

349

Tweedledee and Tweedledoom

SAID the Undertaker to the Overtaker.
Thank you for the butcher and the candlestick-maker.
For the polo player and the pretzel-baker.
For the lawyer and the lover and the wife-forsaker.

Thank you for my bulging, verdant acre.
Said the Undertaker to the Overtaker.
Move in, move under, said the Overtaker.

350 *A Word to Husbands*

TO keep your marriage brimming.
With love in the loving cup,
Whenever you're wrong, admit it;
Whenever you're right, shut up.

351 *What Do You Want: A Meaningful Dialogue, or a Satisfactory Talk?*

BAD money drives out good.
That's Gresham's law, which I have not until recently understood.
No economist I, to economics I have an incurable allergy,
But now I understand Gresham's law through obvious analogy.
Just as bad money drives the good beyond our reach,
So has the jargon of the hippie, the huckster and the bureaucrat debased the sterling of our once lucid speech.
What's worse, it has induced the amnesia by which I am faced;
I can't recall the original phraseology which the jargon has replaced.
Would that I had the memory of a computer or an elephant!
What used I to say instead of uptight, clout and thrust and relevant?
Linguistics becomes an ever eerier area, like I feel like I'm in Oz,
Just trying to tell it like it was.

ROY CAMPBELL

1902–1957

352 *On Some South African Novelists*

YOU praise the firm restraint with which they write—
I'm with you there, of course:
They use the snaffle and the curb all right,
But where's the bloody horse?

353

On the Same

FAR from the vulgar haunts of men
Each sits in her 'successful room',
Housekeeping with her fountain pen
And writing novels with her broom.

LANGSTON HUGHES

1902–1967

354

Little Lyric (of Great Importance)

I WISH the rent
Was heaven sent.

355

Morning After

I WAS so sick last night I
Didn't hardly know my mind.
So sick last night I
Didn't know my mind.
I drunk some bad licker that
Almost made me blind.

Had a dream last night I
Thought I was in hell.
I drempt last night I
Thought I was in hell.
Woke up and looked around me—
Babe, your mouth was open like a well.

I said, Baby! Baby!
Please don't snore so loud.
Baby! Please!
Please don't snore so loud.
You jest a little bit o' woman but you
Sound like a great big crowd.

356 *Wake*

TELL all my mourners
To mourn in red—
Cause there ain't no sense
In my bein' dead.

STEVIE SMITH

1902–1971

357 *On the Death of a German Philosopher*

HE wrote *The I and the It*
He wrote *The It and the Me*
He died at Marienbad
And now we are all at sea.

358 *Mrs Simpkins*

MRS SIMPKINS never had very much to do
So it occurred to her one day that the Trinity wasn't true
Or at least but a garbled version of the truth
And that things had moved very far since the days of her youth.
So she became a spiritualist and at her very first party
Just to give her a feeling of confidence the spirit spoke up hearty:
'Since I crossed over dear friends' it said 'I'm no different to what I was before
Death's not a separation or alteration or parting it's just a one-handled door
We spirits can come back to you if your seance is orthodox
But you can't come over to us till your body's shut in a box
And this is the great thought I want to leave with you today
You've heard it before but in case you forgot death isn't a passing away
It's just a carrying on with friends relations and brightness
Only you don't have to bother with sickness and there's no financial tightness.'
Mrs Simpkins went home and told her husband he was a weak pated fellow
And when he heard the news he turned a daffodil shade of yellow

'What do you mean, Maria?' he cried, 'it can't be true there's no rest
From one's uncles and brothers and sisters nor even the wife of one's
breast?'
'It's the truth,' Mrs Simpkins affirmed, 'there is no separation
There's a great reunion coming for which this life's but a preparation.'
This worked him to such a pitch that he shot himself through the head
And now she has to polish the floors of Westminster County Hall for her
daily bread.

359 *Emily Writes Such a Good Letter*

MABEL was married last week
So now only Tom left

The doctor didn't like Arthur's cough
I have been in bed since Easter

A touch of the old trouble

I am downstairs today
As I write this
I can hear Arthur roaming overhead

He loves to roam
Thank heavens he has plenty of space to roam in

We have seven bedrooms
And an annexe

Which leaves a flat for the chauffeur and his wife

We have much to be thankful for

The new vicar came yesterday
People say he brings a breath of fresh air

He leaves me cold
I do not think he is a gentleman

Yes, I remember Maurice very well
Fancy getting married at his age
She must be a fool

You knew May had moved?
Since Edward died she has been much alone

It was cancer

No, I know nothing of Maud
I never wish to hear her name again
In my opinion Maud
Is an evil woman

Our char has left
And a good riddance too
Wages are very high in Tonbridge

Write and tell me how you are, dear,
And the girls,
Phoebe and Rose
They must be a great comfort to you
Phoebe and Rose.

360 *The Grange*

OH there hasn't been much change
At the Grange,

Of course the blackberries growing closer
Make getting in a bit of a poser,
But there hasn't been much change
At the Grange.

Old Sir Prior died,
They say on the point of leaving for the seaside,
They never found the body, which seemed odd to some
(Not me, seeing as what I seen the butler done.)

Oh there hasn't been much change
At the Grange.

The governess 'as got it now,
Miss Ursy 'aving moved down to the Green Cow—
Proper done out of 'er rights, she was, a b. shame.
And what's that the governess pushes round at nights in the old pram?

Oh there hasn't been much change
At the Grange.

The shops leave supplies at the gate now, meat, groceries,
Mostly old tinned stuff you know from McInnes's,
They wouldn't go up to the door,
Not after what happened to Fred's pa.

Oh there hasn't been much change
At the Grange.

Parssing there early this morning, cor lummy,
I 'ears a whistling sound coming from the old chimney,
Whistling it was fit to bust and not a note wrong,
The old pot, whistling The Death of Nelson.

No there hasn't been much change
At the Grange,

But few goes that way somehow,
Not now.

EVELYN WAUGH

1903–1966

361 from *The Loved One*

[*Lines on the Death of Sir Francis Hinsley*]

They told me, Francis Hinsley, they told me you were hung
With red protruding eye-balls and black protruding tongue.
I wept as I remembered how often you and I
Had laughed about Los Angeles and now 'tis here you'll lie;
Here pickled in formaldehyde and painted like a whore,
Shrimp-pink incorruptible, not lost nor gone before.

ANONYMOUS

362

Three Ghostesses

THREE little ghostesses,
Sitting on postesses,
Eating buttered toastesses,
Greasing their fistesses,
Up to their wristesses.
Oh, what beastesses
To make such feastessess!

WILLIAM PLOMER

1903–1973

363

French Lisette: A Ballad of Maida Vale

WHO strolls so late, for mugs a bait,
In the mists of Maida Vale,
Sauntering past a stucco gate
Fallen, but hardly frail?

You can safely bet that it's French Lisette,
The pearl of Portsdown Square,
On the game she has made her name
And rather more than her share.

In a coat of cony with her passport phony
She left her native haunts,
For an English surname exchanging *her* name
And then took up with a ponce.

Now a meaning look conceals the hook
Some innocent fish will swallow,
Chirping 'Hullo, Darling!' like a cheeky starling
She'll turn, and he will follow,

For her eyes are blue and her eyelids too
And her smile's by no means cryptic,
Her perm's as firm as if waved with glue,
She plies an orange lipstick,

And orange-red is her perky head
Under a hat like a tiny pie—
A pie on a tart, it might be said,
Is redundant, but oh, how spry!

From the distant tundra to snuggle under her
Chin a white fox was conveyed,
And with winks and leerings and Woolworth earrings
She's all set up for trade.

Now who comes here replete with beer?
A quinquagenarian clerk
Who in search of Life has left 'the wife'
And 'the kiddies' in Tufnell Park.

Dear sir, beware! for sex is a snare
And all is not true that allures.
Good sir, come off it! She means to profit
By this little weakness of yours:

Too late for alarm! Exotic charm
Has caught in his gills like a gaff,
He goes to his fate with a hypnotized gait,
The slave of her silvery laugh,

And follows her in to her suite of sin,
Her self-contained bower of bliss,
They enter her flat, she takes his hat,
And he hastens to take a kiss.

Ah, if only he knew that concealed from view
Behind a 'folk-weave' curtain
Is her fancy man, called Dublin Dan,
His manner would be less certain,

His bedroom eyes would express surprise,
His attitude less languor,
He would watch his money, not call her 'Honey',
And be seized with fear or anger.

Of the old technique one need scarcely speak,
But oh, in the quest for Romance
'Tis folly abounding in a strange surrounding
To be divorced from one's pants.

364 *Headline History*

GRAVE CHARGE IN MAYFAIR BATHROOM CASE,
ROMAN REMAINS FOR MIDDLE WEST,
GOLFING BISHOP CALLS FOR PRAYERS,
HOW MURDERED BRIDE WAS DRESSED,

BOXER INSURES HIS JOIE-DE-VIVRE,
DUCHESS DENIES THAT VAMPS ARE VAIN,
DO WOMEN MAKE GOOD WIVES?
GIANT AIRSHIP OVER SPAIN,

SOPRANO SINGS FOR FORTY HOURS,
COCKTAIL BAR ON MOORING MAST,
'NOISE, MORE NOISE!' POET'S LAST WORDS,
COMPULSORY WIRELESS BILL IS PASSED,

ALLEGED LAST TRUMP BLOWN YESTERDAY,
TRAFFIC DROWNS CALL TO QUICK AND DEAD,
CUP TIE CROWD SEES HEAVENS OPE,
'NOT END OF WORLD', SAYS WELL-KNOWN RED.

365 *To the Moon and Back*

countdown takeoff
moonprints rockbox
splashdown claptrap

CYRIL CONNOLLY

1903–1974

366 from *Where Engels Fears to Tread*

[*Thirties Poetry*]

(i)

M IS for Marx
and Movement of Masses
and Massing of Arses
and Clashing of Classes.

(ii)

COME on Percy, my pillion-proud, be
camber-conscious
Cleave to the crown of the road.

(iii)

IT was late last night when my lord came home
enquiring for his lady O
The servants cried on every side
She's gone with the Left Book
Study Circle O!

(iv)

SOMETHING is going to go, baby,
And it won't be your stamp-collection.
Boom!

367 *To Osbert Sitwell*

O OSBERT father Osbert
to whom the young men pray
a Sitwell voice and Sitwell face
bestow on me this day,
and teach me in society
to cleave to what is best
and ogle, flatter, boast, till I
may stink like all the rest.

368

On Geoffrey Grigson

IF art were a series of arbitrary digs,
I'm sure we should all be as merry as grigs.

369

On Himself

AT Eton with Orwell, at Oxford with Waugh,
He was nobody afterwards and nothing before.

DOROTHY FIELDS

1904–1974

370

A Fine Romance

She A FINE romance! with no kisses!
A fine romance, my friend, this is!
We should be like a couple of hot tomatoes,
But you're as cold as yesterday's mashed potatoes.
A fine romance! you won't nestle,
A fine romance, you won't wrestle!
I might as well play bridge with my maiden aunts!
I haven't got a chance.
This is a fine romance!

A fine romance! my good fellow!
You take romance, I'll take Jello!
You're calmer than the seals in the Arctic Ocean,
At least they flap their fins to express emotion.
A fine romance! with no quarrels,
With no insults, and all morals!
I've never mussed the crease in your blue serge pants,
I never get the chance.
This is a fine romance!

He A fine romance! with no kisses!
A fine romance, my friend, this is!

We two should be like clams in a dish of chowder,
But we just fizz like parts of a Seidlitz powder.
A fine romance, with no clinches,
A fine romance, with no pinches,
You're just as hard to land as the 'Île de France'!
I haven't got a chance.
This is a fine romance!

A fine romance! my dear Duchess!
Two old fogies who need crutches!
True love should have the thrills that a healthy crime has!
We don't have half the thrill that the 'March of Time' has!
A fine romance, my good woman!
My strong 'Aged in the Wood' woman!
You never give the orchids I sent a glance!
No! you like cactus plants.
This is a fine romance!

PHYLLIS McGINLEY

1905–1978

371

Evening Musicale

CANDLES. Red tulips, ninety cents the bunch.
Two lions, Grade B. A newly tuned piano.
No cocktails, but a dubious kind of punch,
Lukewarm and weak. A harp and a soprano.
The 'Lullaby' of Brahms. Somebody's cousin
From Forest Hills, addicted to the pun.
Two dozen gentlemen; ladies, three dozen,
Earringed and powdered. Sandwiches at one.

The ash trays few, the ventilation meager.
Shushes to greet the late-arriving guest
Or quell the punch-bowl group. A young man eager
To render 'Danny Deever' by request.
And sixty people trying to relax
On little rented chairs with gilded backs.

372

City Christmas

NOW is the time when the great urban heart
 More warmly beats, exiling melancholy.
Turkey comes table d'hôte or à la carte.
 Our elevator wears a wreath of holly.

Mendicant Santa Claus in flannel robes
 At every counter contradicts his label,
Alms-asking. We've a tree with colored globes
 In our apartment foyer, on a table.

There is a promise—or a threat—of snow
 Noised by the press. We pull our collars tighter.
And twenty thousand doormen hourly grow
 Politer and politer and politer.

373

Village Spa

BY scribbled names on walls, by telephone number,
 Cleft heart, bold slogan, carved in every booth,
This sanctum shall be known. This holy lumber
 Proclaims a temple dedicate to Youth.
Daily in garments lawful to their tribe,
 In moccasins and sweaters, come the Exalted
To lean on spotty counters and imbibe
 Their ritual Cokes or drink a chocolate malted.

This refuge is their own. Here the cracked voice,
 Giving the secret passwords, does not falter.
And here the monstrous deity of their choice
 Sits bellowing from his fantastic altar,
A juke-box god, enshrined and well at home,
 Dreadful with neon, shuddering with chrome.

374

Squeeze Play

JACKSON POLLOCK had a quaint
Way of saying to his sibyl,
'Shall I dribble?
Should I paint?'
And with never an instant's quibble,
Sibyl always answered,
'Dribble.'

375

The Velvet Hand

I CALL that parent rash and wild
Who'd reason with a six-year child,
Believing little twigs are bent
By calm, considered argument.

In bandying words with progeny,
There's no percentage I can see,
And people who, imprudent, do so,
Will wonder how their troubles grew so.

Now underneath this tranquil roof
Where sounder theories have their proof,
Our life is sweet, our infants happy.
In quietude dwell Mammy and Pappy.

We've sworn a stern, parental vow
That argument we won't allow.
Brooking no juvenile excess here,
We say a simple No or Yes, here,

And then, when childish wails begin
We don't debate.
We just give in.

JOHN BETJEMAN

1906–1984

376

A Hike on the Downs

'YES, rub some soap upon your feet!
  We'll hike round Winchester for weeks—
Like ancient Britons—just we two—
  Or more perhaps like ancient Greeks.

'You take your pipe—that will impress
  Your strength on anyone who passes;
I'll take my *Plautus* (*non purgatus*)
  And both my pairs of horn-rimmed glasses.

'I've got my first, and now I know
What life is and what life contains—
For, being just a first year man
You don't meet all the first-class brains.

'Objectively, our Common Room
Is like a small Athenian State—
Except for Lewis: he's all right
But do you think he's *quite* first rate?

'Hampshire mentality is low,
And that is why they stare at us.
Yes, here's the earthwork—but it's dark;
We may as well return by bus.'

377

Hunter Trials

It's awf'lly bad luck on Diana,
Her ponies have swallowed their bits;
She fished down their throats with a spanner
And frightened them all into fits.

So now she's attempting to borrow.
Do lend her some bits, Mummy, *do*;
I'll lend her my own for to-morrow,
But to-day *I*'ll be wanting them too.

Just look at Prunella on Guzzle,
The wizardest pony on earth;
Why doesn't she slacken his muzzle
And tighten the breech in his girth?

I say, Mummy, there's Mrs Geyser
And doesn't she look pretty sick?
I bet it's because Mona Lisa
Was hit on the hock with a brick.

Miss Blewitt says Monica threw it,
But Monica says it was Joan,
And Joan's very thick with Miss Blewitt,
So Monica's sulking alone.

And Margaret failed in her paces,
  Her withers got tied in a noose,
So her coronets caught in the traces
  And now all her fetlocks are loose.

Oh, it's me now. I'm terribly nervous.
  I wonder if Smudges will shy.
She's practically certain to swerve as
  Her Pelham is over one eye.

* * * * * *

Oh wasn't it naughty of Smudges?
  Oh, Mummy, I'm sick with disgust.
She threw me in front of the Judges,
  And my silly old collarbone's bust.

378 *Reproof Deserved*, or *After the Lecture*

WHEN I saw the grapefruit drying, cherry in each centre lying,
  And a dozen guests expected at the table's polished oak,
Then I knew, my lecture finished, I'ld be feeling quite diminished
  Talking on, but unprotected, so that all my spirit broke.

'Have you read the last Charles Morgan?' 'Are you writing for the organ
  Which is published as a vital adjunct to our cultural groups?'
'This year some of us are learning all *The Lady's Not for Burning*
  For a poetry recital we are giving to the troops.'

'Mr Betjeman, I grovel before critics of the novel,
  Tell me, if I don't offend you, have you written one yourself?
You haven't? Then the one I wrote is (not that I expect a notice)
  Something I would like to send you, just for keeping on your shelf.'

'Betjeman, I bet your racket brings you in a pretty packet
  Raising the old lecture curtain, writing titbits here and there.
But, by Jove, your hair is thinner, since you came to us in Pinner,
  And you're fatter now, I'm certain. What you need is country air.'

This and that way conversation, till I turn in desperation
  To a kind face (can I doubt it?) mercifully mute so far.
'Oh,' it says, 'I missed the lecture, wasn't it on architecture?
  Do please tell me all about it, what you do and who you are.'

379 *Longfellow's Visit to Venice*

(To be read in a quiet New England accent)

NEAR the celebrated Lido where the breeze is fresh and free
Stands the ancient port of Venice called the City of the Sea.

All its streets are made of water, all its homes are brick and stone,
Yet it has a picturesqueness which is justly all its own.

Here for centuries have artists come to see the vistas quaint,
Here Bellini set his easel, here he taught his School to paint.

Here the youthful Giorgione gazed upon the domes and towers,
And interpreted his era in a way which pleases ours.

A later artist, Tintoretto, also did his paintings here,
Massive works which generations have continued to revere.

Still to-day come modern artists to portray the buildings fair
And their pictures may be purchased on San Marco's famous Square.

When the bell notes from the belfries and the campaniles chime
Still to-day we find Venetians elegantly killing time

In their gilded old palazzos, while the music in our ears
Is the distant band at Florians mixed with songs of gondoliers.

Thus the New World meets the Old World and the sentiments expressed
Are melodiously mingled in my warm New England breast.

380 *The Old Land Dog*

(After Henry Newbolt)

OLD General Artichoke lay bloated on his bed,
Just like the Fighting Téméraire.
Twelve responsive daughters were gathered round his head
And each of them was ten foot square.

Old General Artichoke he didn't want to die:
He never understood the truth and that perhaps was why
It wouldn't be correct to say he always told a lie.
Womenfolk of England, oh beware!

'Fetch me down my rifle—it is hanging in the hall'
Just like the Fighting Téméraire;
'Lydia, get my cartridge cases, twenty-four in all',
And each of them is ten foot square.

'I'll tell you all in detail, girls, my every campaign
In Tuscany, Bolivia, Baluchistan and Spain;
And when I've finished telling you, I'll tell you all again';
Womenfolk of England, oh beware!

Old General Artichoke he's over eighty-two,
Just like the Fighting Téméraire.
His daughters all make rush mats when they've nothing else to do,
And each of them is ten foot square.

Now all ye pension'd army men from Tunbridge Wells to Perth,
Here's to General Artichoke, the purplest man on earth!
Give three loud cheers for Cheltenham, the city of his birth.
Womenfolk of England, oh beware!

381 *The Ballad of George R. Sims*

IT'S an easy game, this reviewin'—the editor sends yer a book,
Yer puts it down on yer table and yer gives it a 'asty look,
An' then, Sir, yer writes about it as though yer 'ad read it all through,
And if ye're a pal o' the author yer gives it a good review.

But if the author's a wrong 'un—and *some* are, as I've 'eard tell—
Or if 'e's a stranger to yer, why then yer can give him 'ell.
So what would yer 'ave me do, Sir, to humour an editor's whims,
When I'm pally with Calder-Marshall, and never knew George R. Sims!

reviewin'] this was a verse review in the *New Statesman* (25 Oct. 1968) of a selection of George R. Sims's Ballads introduced by Arthur Calder-Marshall

It is easy for you to deride me and brush me off with a laugh
And say 'Well, the answer's potty—yer review it just 'arf and 'arf'—
For I fear I must change my tune, Sir, and pump the bellows of praise
And say that both 'alves are good, Sir, in utterly different ways.

I'm forgettin' my cockney lingo—for I lapse in my style now and then
As Sims used to do in his ballads when he wrote of the Upper Ten—
'Round in the sensuous galop the high-born maids are swung
Clasped in the arms of *roués* whose vice is on every tongue'.

'It was Christmas Day in the workhouse' is his best known line of all,
And this is his usual metre, which comes, as you may recall,
Through Tennyson, Gordon, Kipling and on to the Sergeants' Mess,
A rhythm that's made to recite in, be it mufti or evening dress.

Now Arthur shows in his intro that George R. Sims was a bloke
Who didn't compose his ballads as a sort of caustic joke;
He cared about social justice but he didn't aim very high
Though he knew how to lay on the sobstuff and make his audience cry.

The village church on the back-drop is painted over for good,
The village concerts are done for where the Young Reciter stood,
The magic-lantern is broken and we laugh at the mission hymns—
We laugh and we well might weep with the Ballads of George R. Sims.

LUIS d'ANTIN VAN ROOTEN

1906–1973

382 from *Mots d'Heures: Gousses, Rames*

Author's note:

'Words of the Hours.' A more poetic title than the more familiar 'Book of Hours.' A religious or philosophical background is tacitly indicated by this title.

'Gousses, Rames.' A 'gousse' is a clove or section, as in the bulb of the garlic plant. We can therefore assume that this implies 'Root and Branch,' or a complete unity. Alas, would only that the poems had come down to us so.

(i)

JACQUES s'apprête
Coulis de nos fêtes.[1]
Et soif que dites nos lignes.[2]
Et ne sauve bédouine tempo[3] y aussi,
Telle y que de plat terre, cligne.[4]

[1] *Coulis*: a sort of strained broth. Jacques was either a sauce chef or an invalid.
[2] Jacques was also an alcoholic, since his thirst is beyond description.
[3] He was fond of Arab music.
[4] He believed the earth was flat. The last word of the line, meaning 'wink,' is obviously a stage direction. Poor Jacques, whoever he was, was obviously considered a fool.

(ii)

LIT-ELLE messe, moffette,[1]
Satan ne te fête,
Et digne somme cœurs et nouez.
À longue qu'aime est-ce pailles d'Eure.
Et ne Satan bise ailleurs
Et ne fredonne messe. Moffette, ah, ouais![2]

[1] *Moffette*. noxious exhalations formed in underground galleries or mines.
[2] This little fragment is a moral precept addressed to a young girl. She is advised to go to mass even under the most adverse conditions in order to confound Satan and keep her heart pure until the knot (marriage) is tied. She is warned against long engagements and to stay out of hayfields, be they as lush and lovely as those of the Eure valley, for Satan will not be off spoiling crops elsewhere. She must not mumble at mass, or the consequences will make the noxious fumes of earth seem trivial.

(iii)

POLIS poutre catalane
Polis poutre catalane
Polis poutre catalane
En la sève ti.[1]
Sou quitté qu'étoffe à gain
Sou quitté qu'étoffe à gain
Sou quitté qu'étoffe à gain
Des vols gagne où est.[2]

[1] The repetition of the first line to form a quatrain would indicate that this was a song or perhaps a children's game chant. The first verse advises the use of Ti sap to polish a Catalan beam. The Ti plant is common to the South Seas but heretofore has had no virtue attributed to it other than its decorative value.
[2] 'A penny taken is the stuff of profit | There is gain in stealing.' A far cry from 'A penny saved is a penny earned.' If the first four lines are considered a derision of painstaking labor carried to exotic lengths, the second four clearly make this a thieves' refrain.

(iv)

'POUSSE y gâte, pousse y gâte,
Et Arabe, yeux bine?'[1]
'A ben, tout l'on donne
Toluca de couenne.'[2]
'Pousse y gâte, pousse y gâte,
Oh, a dit Dieu d'hère?'[3]
'Y fraternelle Lydie, Moïse,
Honneur de chair.'[4]

[1] Although the dialogue form of versifying is very ancient and quite common, this is one of the few fragments so written. In the first speech, an Arab is chided for planting a crop, then allowing it to spoil, while merely eyeing his hoe. The Arabs are a traditionally nomadic people, not given to agriculture.

[2] In his reply, our hero admits he was building castles in Spain, dreaming of a pigskin from Toluca (famous market town, capital of the State of Mexico, Mexico). These pigskins make excellent water bags—an item of great interest to a desert-dweller.

[3] 'What sayeth the poor man's God?' *Hère*: in medieval times, a serf, bound to the lord of a manor.

[4] 'There will be brotherhood in Lydia, Moses, blood is thicker than water.' Lydia: Middle East region bordering on the Aegean Sea.

ANONYMOUS

383

Scones

I ASKED the maid in dulcet tone
To order me a buttered scone.
The silly girl has been and gone
And ordered me a buttered scone.

WILLIAM EMPSON

1906–1984

384

Just a Smack at Auden

WAITING for the end, boys, waiting for the end.
What is there to be or do?
What's to become of me or you?
Are we kind or are we true?
Sitting two and two, boys, waiting for the end.

Shall I build a tower, boys, knowing it will rend,
Crack upon the hour, boys, waiting for the end?
Shall I pluck a flower, boys, shall I save or spend?
All turns sour, boys, waiting for the end.

Shall I send a wire, boys? Where is there to send?
All are under fire, boys, waiting for the end.
Shall I turn a sire, boys? Shall I choose a friend?
The fat is in the pyre, boys, waiting for the end.

Shall I make it clear, boys, for all to apprehend,
Those that will not hear, boys, waiting for the end,
Knowing it is near, boys, trying to pretend,
Sitting in cold fear, boys, waiting for the end?

Shall we send a cable, boys, accurately penned,
Knowing we are able, boys, waiting for the end,
Via the Tower of Babel, boys? Christ will not ascend.
He's hiding in his stable, boys, waiting for the end.

Shall we blow a bubble, boys, glittering to distend,
Hiding from our trouble, boys, waiting for the end?
When you build on rubble, boys, Nature will append
Double and re-double, boys, waiting for the end.

Shall we make a tale, boys, that things are sure to mend,
Playing bluff and hale, boys, waiting for the end?
It will be born stale, boys, stinking to offend,
Dying ere it fail, boys, waiting for the end.

Shall we all go wild, boys, waste and make them lend,
Playing at the child, boys, waiting for the end?
It has all been filed, boys, history has a trend,
Each of us enisled, boys, waiting for the end.

What was said by Marx, boys, what did he perpend?
No good being sparks, boys, waiting for the end.
Treason of the clerks, boys, curtains that descend,
Lights becoming darks, boys, waiting for the end.

Waiting for the end, boys, waiting for the end.
Not a chance of blend, boys, things have got to tend.
Think of those who vend, boys, think of how we wend,
Waiting for the end, boys, waiting for the end.

JOHN SPARROW

1906–1992

385

To an Angel in the House

A PAT on the head
Sends me happy to bed,
I wish—how I wish—you'd do that more!
Cold words and neglect
Leave me wretched and wrecked:
Don't send me to Coventry—pat more!

386

Epitaph

THIS stone, with not unpardonable pride,
Proves by its record what the world denied:
Nemo could do a natural thing—he died.

387

Apology and Explanation

I'M sorry I'm late,
But it's always my fate,
So don't get excited, or shout;
A dear little negress
Obstructed the egress—
It took me an age to get out.

And then, on the way,
There was further delay
Of a kind that I couldn't foresee:
A horrible ogress
Impeded my progress—
Why must these things happen to me,
Poor me?
Why *must* these things happen to me?

W. H. AUDEN

1907–1973

388

Statesmen

WHEN statesmen gravely say, 'We must be realistic',
The chances are they're weak and therefore pacifistic:
But when they speak of Principles—look out—perhaps
Their generals are already poring over maps.

389

Passenger Shanty

THE ship weighed twenty thousand ton
    Parlez-vous
The ship weighed twenty thousand ton
    Parlez-vous
She left Marseille at a quarter-to-one
For the China War and the tropical sun.
    Inky-pinky-parlez-vouz.

The passengers are rather *triste*,
There's many a fool, and many a beast,
Who ought to go west, but is bound for the East.

Mr Jackson buys rubber and sells it again,
He paints in oils and he drinks champagne,
Says: 'I should have been born in Elizabeth's reign.'

His wife learns astrology out of a book,
Says: 'Your horoscope's queer and I don't like its look.
With the Moon against Virgo you might be a crook.'

The planter tells us: 'In Malay
We play rugger in March and cricket in May
But feel starved for sex at the end of the day.'

The journalist Capa plays dicing games,
He photographed Teruel Town in flames,
He pinches the bottoms of all the dames.

The Dominican monks get up with the sun,
They're as fond of their dinner as anyone,
And they have their own mysterious fun.

The belle of the boat-deck laughs like a jay,
She models her face upon *Ta Beauté*
And her eyebrows are shifted every day.

Her rival, the bitch from Aix-les-Bains,
Has a Pekingese *nez* and monkey's *mains*
And her *buste*, it would seem, has been flattened by *trains*.

Alphonse is a student who talks a lot,
When he walks upstairs, he waggles his bot.
He says: 'France may go left, but Annam must not.'

The idiot child stole a cigarette;
The father looks at his wife with regret
And thinks of that night at the *Bal Musette*.

The Siamese doctor is plump and tan,
He thinks shouting a mark of the Westernised Man.
But he cures VD—which is more than you can.

The beautiful *matelots* and *mousses*
Would be no disgrace to the Ballets Russes,
But I can't see their presence is very much use.

It's the engineers who run the ship
And the stewards who bring us our tots to sip
Without which we'd never get through the trip.

Christopher sends off letters by air,
He longs for Someone who isn't there,
But Wystan says: 'Love is exceedingly rare.'

The sea is *blau*, the sea is *tief*
    Parlez-vous
The sea is *blau*, the sea is *tief*
    Parlez-vous
C'est le cimetière du Château d'If.
No doubt. But it's dull beyond belief.
    Inky-pinky-parlez vous.

390

Give me a doctor . . .

GIVE me a doctor partridge-plump,
Short in the leg and broad in the rump,
An endomorph with gentle hands
Who'll never make absurd demands
That I abandon all my vices
Nor pull a long face in a crisis,
But with a twinkle in his eye
Will tell me that I have to die.

391

The Love Feast

IN an upper room at midnight
See us gathered on behalf
Of love according to the gospel
Of the radio-phonograph.

Louis telling Anne what Molly
Said to Mark behind her back;
Jack likes Jill who worships George
Who has the hots for Jack.

Catechumens make their entrance;
Steep enthusiastic eyes
Flicker after tits and baskets;
Someone vomits; someone cries.

Willy cannot bear his father,
Lilian is afraid of kids;
The Love that rules the sun and stars
Permits what He forbids.

Adrian's pleasure-loving dachshund
In a sinner's lap lies curled;
Drunken absent-minded fingers
Pat a sinless world.

390 *endomorph*] a person of rounded build

391 *Catechumens*] candidates for baptism

Who is Jenny lying to
By long-distance telephone?
The Love that made her out of nothing
Tells me to go home.

But that Miss Number in the corner
Playing hard to get. . . .
I am sorry I'm not sorry . . .
Make me chaste, Lord, but not yet.

392

The Aesthetic Point of View

As the poets have mournfully sung,
Death takes the innocent young,
The rolling-in-money,
The screamingly-funny,
And those who are very well hung.

LOUIS MacNEICE

1907–1963

393

from *Autumn Journal*

I OUGHT to be glad
That I studied the classics at Marlborough and Merton,
Not everyone here having had
The privilege of learning a language
That is incontrovertibly dead,
And of carting a toy-box of hall-marked marmoreal phrases
Around in his head.
We wrote compositions in Greek which they said was a lesson
In logic and good for the brain;
We marched, counter-marched to the field-marshal's blue-pencil baton,
We dressed by the right and we wrote out the sentence again.
We learned that a gentleman never misplaces his accents,
That nobody knows how to speak, much less how to write
English who has not hob-nobbed with the great-grandparents of English,
That the boy on the Modern Side is merely a parasite

But the classical student is bred to the purple, his training in syntax
  Is also a training in thought
And even in morals; if called to the bar or the barracks
  He always will do what he ought.
And knowledge, besides, should be prized for the sake of knowledge:
  Oxford crowded the mantelpiece with gods—
Scaliger, Heinsius, Dindorf, Bentley and Wilamowitz—
  As we learned our genuflexions for Honour Mods.
And then they taught us philosophy, logic and metaphysics,
  The Negative Judgment and the Ding an Sich,
And every single thinker was powerful as Napoleon
  And crafty as Metternich.
And it really was very attractive to be able to talk about tables
  And to ask if the table *is*,
And to draw the cork out of an old conundrum
  And watch the paradoxes fizz.
And it made one confident to think that nothing
  Really was what it seemed under the sun,
That the actual was not real and the real was not with us
  And all that mattered was the One.
And they said 'The man in the street is so naïve, he never
  Can see the wood for the trees;
He thinks he knows he sees a thing but cannot
  Tell you how he knows the thing he thinks he sees.'
And oh how much I liked the Concrete Universal,
  I never thought that I should
Be telling them vice-versa
  That they can't see the trees for the wood.
But certainly it was fun while it lasted
  And I got my honours degree
And was stamped as a person of intelligence and culture
  For ever wherever two or three
Persons of intelligence and culture
  Are gathered together in talk
Writing definitions on invisible blackboards
  In non-existent chalk.
But such sacramental occasions
  Are nowadays comparatively rare;
There is always a wife or a boss or a dun or a client
  Disturbing the air.
Barbarians always, life in the particular always,
  Dozens of men in the street,
And the perennial if unimportant problem
  Of getting enough to eat.

So blow the bugles over the metaphysicians,
  Let the pure mind return to the Pure Mind;
I must be content to remain in the world of Appearance
  And sit on the mere appearance of a behind.
But in case you should think my education was wasted
  I hasten to explain
That having once been to the University of Oxford
  You can never really again
Believe anything that anyone says and that of course in an asset
  In a world like ours;
Why bother to water a garden
  That is planted with paper flowers?
O the Freedom of the Press, the Late Night Final,
  To-morrow's pulp;
One should not gulp one's port but as it isn't
  Port, I'll gulp it if I want to gulp
But probably I'll just enjoy the colour
  And pour it down the sink
For I don't call advertisement a statement
  Or any quack medicine a drink.
Good-bye now, Plato and Hegel,
  The shop is closing down;
They don't want any philosopher-kings in England,
  There ain't no universals in this man's town.

NICHOLAS BENTLEY

1907–1978

394

Cecil B. De Mille

CECIL B. DE MILLE,
Rather against his will,
Was persuaded to leave Moses
Out of *The Wars of the Roses*.

395

The Londonderry Air

YOU can tell by the angle
Of Lord Londonderry's hat
That *he* is not a member
Of the proletariat.
And in case this may escape you,
A reminder of it shows
Also very clearly
In the angle of his nose.

396

On Lady A——

GATHER ye rosebuds while ye may,
  Old Time is still a-flying,
Despite the pads, the whale-bone stay,
  The rougeing and the dyeing.

That glorious lamp of heaven, the moon,
  Is usually a-setting
While to Cole Porter's latest tune
  Ye still are seen curvetting.

That age is best which is the first,
  When looks and limbs are younger;
'Tis past recall, although ye thirst
  And amorously hunger.

A. D. HOPE

1907–

397

Möbius Strip-Tease

AN erudite demon, a fiend in topology,
Shaped much like a grin on a sphere on a trivet,
To add to the carnal advancement of knowledge he
Invented a woman. Now, would you believe it?

A woman so modelled no man could resist her,
So luscious her curves, so alluring her smile,
Yet no daughter of Eve's could claim her for sister,
Though equally formed to seduce and beguile.

For her surface—a pure aphrodisiac plastic—
No mathematician could ever equate
By any contortion or motion elastic
To those we caress in man's fallen estate.

O she was a heartache! O she was a honey!
The fiend asked his friends gathered round in a ring:
'A degenerate set! Would you bet even money,
Though she looks like a succubus fit for a King?'

'Come off it,' they answered, 'her shape is a woman's,
So she can't be a true topological freak,
Though a singleton, maybe, to ordinary humans
Who think any girl they adore is unique.'

'In our rubber sheet world,' said the fiend with a chortle
Converting himself to a three-masted barque,
'Equivalent shapes may delude a poor mortal,
But *you* should know Woman's distinguishing mark.'

'A woman's a man-trap,' they answered in chorus,
'A trochus with trunnions, a tunnel to Hell;
Reduced to essentials she's simply a torus
And this must apply to your temptress as well.'

'Alas, my poor friends you are sadly mistaken:
This exquisite creature is built to deceive;
For the Devil's own cunning will not save his bacon
When caught in the nets that topologists weave.

'This marvellous manifold's not like a doughnut,
Quoit or cat's-cradle or twists of red tape,
And though very tortive, she screws like no known nut;
So I'd better explain her remarkable shape.

'Like a Boy Surface girl, my delightful invention
In Euclidean space is too awkward to plot,
But in Hell, with the help of an extra dimension
And a regressive cut, she's a true-lovers'-knot,

'Though she looks like a woman from thrutch-piece to throttle,
If you follow my clew of a Möbius strip-tease,
She is really a camouflaged double Klein bottle
With only one surface unlike other shes.

'Four Möbius strips brought my plan to fruition,
Ingeniously joined by original sin;
If you rise to the urgings of male intuition,
You'll find yourself out every time you go in.

'She cannot be mated or orientated,
Nor is homeomorphic to any known male;
And though in her arms you may feel quite elated,
All further advances are destined to fail.

'And before we proceed to our first Demon-stration,
May I venture to say, with excusable pride,
That this elegant essay in total frustration
Justifies mathematics, both pure and applied.

'Furthermore, as a torment for sinful seducers,
I think I may claim for the very first time,
To have added to Hell's repertoire something new, sirs:
A case where the punishment *won't* fit the crime.'

Glossary for Non-Mathematical Demons

Topology — A field of Botany invaded by certain mathematicians with a sense of humour; devoted to studying the shapes of things.

Möbius, Klein & Boy — Topologists of great eminence and a profound sense of humour.

Elastic motion — The imaginary shift of spatial points required to change one spatial shape to a mathematically equivalent shape. Also something girls do without any mathematical knowledge at all.

Degenerate set — A coarse logical term for a class of things containing only one member; a member of the class of classes of unique individuals; a mathematical term of abuse.

Succubus — A theological entity, rather than a mathematical one; if you don't know what it is, you'd better not worry your pretty little head about it.

Singleton — See 'degenerate set': nothing to do with bridge.

Rubber sheet world	Topology (for topologists), otherwise something out of Grimm to help Frog Princes to bed.
Trochus	Anything in the shape of a wheel; in topology it might be a lot of other things as well.
Trunnions	Arms (or legs) of a cannon barrel.
Torus	A refined (or mathematical) word for anything shaped like a doughnut.
Manifold	A connected surface such that if you caress it, it will respond by being thigmotactic to your hand—such as a girl or a football.
Tortive	Twisty or twistable, according to your intentions.
Boy surface	A very sophisticated three-dimensional figure with only one surface. Invented by Mr Boy (see Möbius, Klein etc.).
Regressive cut	A mathematical way of getting your own back and making some surprising discoveries on the way.
Thrutch-piece	Consult a very big dictionary; it probably won't help you, but your imagination may.
Möbius strip	What happens when you twist your belt putting it on. It has only one side and one edge and numerous even more remarkable properties.
Klein bottle	An attempt to make two Möbius strips copulate without benefit of more than three dimensions; a hell of a topological joke.
Original sin	Not a mathematical operation as far as can be proved—but you never know.
Male intuition	Not a mathematical idea either, but it has associations with binary arithmetic.
Orientated	You wouldn't understand this anyway—a topological technicality.
Homeomorphic	Topologically equivalent in shape.
Demon-stration	Just an ordinary demo, but conducted in another place.
The crime	See 'Original sin'.

PATRICK BARRINGTON

1908–1990

398 *I Had a Duck-Billed Platypus*

I HAD a duck-billed platypus when I was up at Trinity,
With whom I soon discovered a remarkable affinity.
He used to live in lodgings with myself and Arthur Purvis,
And we all went up together for the Diplomatic Service.
I had a certain confidence, I own, in his ability,
He mastered all the subjects with remarkable facility;
And Purvis, though more dubious, agreed that he was clever,
But no one else imagined he had any chance whatever.
I failed to pass the interview, the Board with wry grimaces
Took exception to my boots and then objected to my braces,
And Purvis too was failed by an intolerant examiner
Who said he had his doubts as to his sock-suspenders' stamina.
The bitterness of failure was considerably mollified,
However, by the ease with which our platypus had qualified.
The wisdom of the choice, it soon appeared, was undeniable;
There never was a diplomat more thoroughly reliable.
He never made rash statements his enemies might hold him to,
He never stated anything, for no one ever told him to,
And soon he was appointed, so correct was his behaviour,
Our Minister (without Portfolio) to Trans-Moravia.
My friend was loved and honoured from the Andes to Esthonia,
He soon achieved a pact between Peru and Patagonia,
He never vexed the Russians nor offended the Rumanians,
He pacified the Letts and yet appeased the Lithuanians,
Won approval from his masters down in Downing Street so wholly, O,
He was soon to be rewarded with the grant of a Portfolio.

When, on the Anniversary of Greek Emancipation,
Alas! He laid an egg in the Bulgarian Legation.
This untoward occurrence caused unheard-of repercussions,
Giving rise to epidemics of sword-clanking in the Prussians.
The Poles began to threaten, and the Finns began to flap at him,
Directing all the blame for this unfortunate mishap at him;
While the Swedes withdrew entirely from the Anglo-Saxon dailies
The right of photographing the Aurora Borealis,
And, all efforts at rapprochement in the meantime proving barren,
The Japanese in self-defence annexed the Isle of Arran.

My platypus, once thought to be more cautious and more tentative
Than any other living diplomatic representative,
Was now a sort of warning to all diplomatic students
Of the risks attached to negligence, the perils of imprudence,
And, branded in the Honours List as 'Platypus, Dame Vera,'
Retired, a lonely figure, to lay eggs at Bordighera.

399 *Take me in your Arms, Miss Moneypenny-Wilson*

TAKE me in your arms, Miss Moneypenny-Wilson,
   Take me in your arms, Miss Bates;
Fatal are your charms, Miss Moneypenny-Wilson,
   Fatal are your charms, Miss Bates;
Say you are my own, Miss Moneypenny-Wilson,
   Say you are my own, Miss Bates;
You I love alone, Miss Moneypenny-Wilson,
   You, and you alone, Miss Bates.

Sweet is the morn, Miss Moneypenny-Wilson;
   Sweet is the dawn, Miss B.,
But sweeter than the dawn and the daisies on the lawn
   Are you, sweet nymphs, to me.
Sweet, sweet, sweet is the sugar to the beet,
   Sweet is the honey to the bee,
But sweeter far than such sweets are
   Are your sweet names to me.

Deaf to my cries, Miss Moneypenny-Wilson,
   Deaf to my sighs, Miss B.,
Deaf to my songs and the story of my wrongs,
   Deaf to my minstrelsy;
Deafer than the newt to the sound of a flute,
   Deafer than a stone to the sea;
Deafer than a heifer to the sighing of a zephyr
   Are your deaf ears to me.

Cold, cold, cold as the melancholy mould,
   Cold as the foam-cold sea,
Colder than the shoulder of a neolithic boulder
   Are the shoulders you show to me.
Cruel, cruel, cruel is the flame to the fuel,
   Cruel is the axe to the tree,
But crueller and keener than a coster's concertina
   Is your cruel, cruel scorn to me.

THEODORE ROETHKE

1908–1963

400

Pipling

BEHOLD the critic, pitched like the *castrati*,
Imperious youngling, though approaching forty;
He heaps few honors on a living head;
He loves himself, and the illustrious dead;
He pipes, he squeaks, he quivers through his nose,—
Some cannot praise him: *I* am one of those.

401

The Mistake

HE left his pants upon a chair:
She was a widow, so she said:
But he was apprehended, bare,
By one who rose up from the dead.

402

Duet

She O WHEN you were little, you were really big:
Now you run to the money, it's jig, jig, jig;
You're becoming that horror, a two-legged pig
Both —In spite of Soren Kierkegaard.

He I'll face all that, and the Divine Absurd:
You be an adverb, I'll be a verb,
I'll spit over my chin and beyond the curb,
Both —And close up that chapter of Kierkegaard.

She We'll sail away from the frightful shore
Of multiple choice and Either/or
To the land where the innocent stretch and snore
Both —With never a thought for Kierkegaard.

She I'm shanty Irish
He —And *pissoir* French?
She I'm a roaring girl, an expensive wench,
Both But at least we know one needn't blench
—In fear and trembling, dear Kierkegaard.

She A mistress of Zen, I'll bite your thumb,
I'll jump on your belly, I'll kick your bum
Till you come to the land of Kingdom Come
Both —Far beyond, O Beyond! dear Kierkegaard.

He My jug, my honey, my can of beer,
She My ex-existentialist darling dear,
Both Should Dame Anxiety ever come near
We'll give each other a box on the ear,
—In honor of Father Kierkegaard.

ROBERT GARIOCH
(Robert Sutherland)

1909–1981

403

Did ye See me?

I'LL tell ye of ane great occasioun:
I tuke pairt in a graund receptioun.
Ye cannae hae the least perceptioun
hou pleased I was to get the invitatioun

tae assist at ane dedicatioun.
And richtlie sae; frae its inceptioun
the hale ploy was my ain conceptioun;
I was asked to gie a dissertatioun.

The functioun was held in the aipen air,
a peety, that; the keelies of the toun,
a toozie lot, gat word of the affair.

We cudnae stop it: they jist gaithert roun
to mak sarcastic cracks and grin and stare.
I wisht I hadnae worn my MA goun.

ploy] sport, frolic *keelies*] louts *toozie*] rough

404

I Was Fair Beat

I SPENT a nicht amang the cognoscenti,
a hie-brou clan, ilk wi a beard on him
like Mark Twain's miners, due to hae a trim,
their years on aiverage roun three-and-twenty.

Of poetry and music we had plenty,
owre muckle, but ye maun be in the swim:
Kurt Schwitters' Ur-sonata that gaes 'Grimm
glimm gnimm bimmbimm,' it fairly wad hae sent ye

daft, if ye'd been there; modern jazz wi juicy
snell wud-wind chords, three new yins, I heard say
by thaim that ken't, new, that is, sen Debussy.

Man, it was awfie. I wad raither hae
a serenata sung by randy pussy,
and what a time a reel of tape can play!

405

A Fair Cop

CASTALIAN Scots, nou may ye cry, Allace!
sen your True Rhymer, Garioch, met a leddy
polis, maist unexpeckit, in a shady
neuk near Tollcross, and nou he's in disgrace.

She met him, raither, but in onie case
it makes nae odds; she got her notebuik ready,
lickit her keelivine, and, jeez! she said he
wes 'urinating in a public place.'

Her very words, a richt wee caution! Pray
forgie me thon expression in nine letters,
a terrible expense: seevin wad dae.

Fending us aa frae muggers and sic craiturs,
the Polis are maist eydent, I daursay,
but fancy fashin wi sic piddlan maitters!

leddy polis] policewoman *keelivine*] pencil *eydent*] eager

MALCOLM LOWRY

1909–1957

406

Epitaph

MALCOLM LOWRY
Late of the Bowery
His prose was flowery
And often glowery
He lived, nightly, and drank, daily,
And died playing the ukulele.

JOHNNY MERCER

1909–1976

407

Jubilation T. Cornpone

WHEN we fought the Yankees and annihilation was near,
Who was there to lead the charge that took us safe to the rear?

Why it wuz Jubilation T. Cornpone,
Old 'Toot-Your-Own-Horn-pone,'
Jubilation T. Cornpone,
A man who knew no fear.

When we almost had 'em but the issue still was in doubt,
Who suggested the retreat that turned it into a rout?

Why it was Jubilation T. Cornpone,
Old 'Tattered-and-Torn-pone,'
Jubilation T. Cornpone,
Who kept us hidin' out.

With our ammunition gone and faced with utter defeat,
Who was it that burned the crops and left us nothin' to eat?

Why it wuz Jubilation T. Cornpone,
Old 'September-Morn-pone,'
Jubilation T. Cornpone,
The pants blown off his seat.

W. R. RODGERS

1909–1969

408 *Home Thoughts from Abroad*

HEARING, this June day, the thin thunder
Of far-off invective and old denunciation
Lambasting and Lambegging the homeland,
I think of that brave man Paisley, eyeless
In Gaza, with a daisy-chain of millstones
Round his neck; groping, like blind Samson,
For the soapy pillars and greased poles of lightning
To pull them down in rains and borborygmic roars
Of rhetoric. (There but for the grace of God
Goes God.) I like his people and I like his guts
But I dislike his gods who always end
In gun-play. Some day, of course, he'll be one
With the old giants of Ireland—such as
Denis of the Drought, or Iron-Buttocks—
Who had at last to be reduced to size,
Quietly shrunken into 'wee people'
And put out to grass on the hills for good,
Minimized like cars or skirts or mums;
Photostatted to fit a literate age
And filed safely away on the dark shelves
Of memory; preserved in ink, oak-gall,
Alcohol, aspic, piety, wit. A pity,
Perhaps, if it is drama one wants. But,
Look at it this way: in this day and age
We can't really have giants lumbering
All over the place, cluttering it up,
With hair like ropes, flutes like telegraph poles,

Home Thoughts] Rodgers was an Ulsterman

And feet like tramcars, intent only on dogging
The fled horse of history and the Boyne.
So today across the Irish Sea I wave
And wish him well from the bottom of my heart
Where truth lies bleeding, its ear-drums burst
By the blather of his hand-me-down talk.
In fond memory of his last stand
I dedicate this contraceptive pill
Of poetry to his unborn followers,
And I place
This bunch of beget-me-nots on his grave.

JOYCE GRENFELL

1910–1979

409 *Stately as a Galleon*

My neighbour, Mrs Fanshaw, is portly-plump and gay,
She must be over sixty-seven, if she is a day.
You might have thought her life was dull,
It's one long whirl instead.
I asked her all about it, and this is what she said:

I've joined an Olde Thyme Dance Club, the trouble is that there
Are too many ladies over, and no gentlemen to spare.
It seems a shame, it's not the same,
But still it has to be,
Some ladies have to dance together,
One of them is me.

Stately as a galleon, I sail across the floor,
Doing the Military Two-step, as in the days of yore.
I dance with Mrs Tiverton; she's light on her feet, in spite
Of turning the scale at fourteen stone, and being of medium height.
So gay the band,
So giddy the sight,
Full evening dress is a must,
But the zest goes out of a beautiful waltz
When you dance it bust to bust.

So, stately as two galleons, we sail across the floor,
Doing the Valse Valeta as in the days of yore.
The gent is Mrs Tiverton, I am her lady fair,
She bows to me ever so nicely and I curtsey to her with care.
So gay the band,
So giddy the sight,
But it's not the same in the end
For a lady is never a gentleman, though
She may be your bosom friend.

So, stately as a galleon, I sail across the floor,
Doing the dear old Lancers, as in the days of yore.
I'm led by Mrs Tiverton, she swings me round and round
And though she manoeuvres me wonderfully well
I never get off the ground.
So gay the band,
So giddy the sight,
I try not to get depressed.
And it's done me a power of good to explode,
And get this lot off my chest.

PETER DE VRIES

1910–1993

410 *Sacred and Profane Love,* or *There's Nothing New under the Moon Either*

WHEN bored by the drone of the wedlocked pair,
When bromides of marriage have started to wear,
Contemplate those of the crimson affair:
  'I had to see you,' and 'Tonight belongs to us.'

Skewered on bliss of a dubious sort
Are all adventurers moved to consort
With others inspiring this hackneyed retort:
  'I can't fight you any longer.'

Some with such wheezes have gone to the dead,
Oblivious that *Liebestod* lurked up ahead,
That pistols would perforate them as they said:
  'This thing is bigger than both of us.'

Experimentation in matters of sin
Pales on the instant it's destined to win;
Paramours end as conformers begin:
  'I don't want just this—I want *you*.'

Explorers are highly unlikely to hear
Novelties murmured into their ear;
Checkered with such is the checkered career:
  'It's not you I'm afraid of, it's myself.'

Such liturgies standardize lovers in league
That someone will cry in the midst of intrigue,
And someone will hear in the midst of fatigue:
  'You don't want *me*—you just want sex!'

Strait is the gate and narrow the way
Closing at last on the ranging roué;
Who plucks a primrose plants a cliché:
  'We're married in the eyes of Heaven.'

The dangerous life is so swiftly prosaic
You might as well marry and live in Passaic;
It ends and begins in established mosaic:
  'I'm all mixed up.'

The lexicon's written for groom and for rake.
Liaisons are always a give-and-take.
Disillusionment's certain to follow a break.
  'For God's sake be careful or someone will hear you!'

RICHARD USBORNE

1910–

411

Epitaph on a Party Girl

LOVELY Pamela, who found
One sure way to get around
Goes to bed beneath this stone
Early, sober, and alone.

STANLEY J. SHARPLESS

1910–

412

Low Church

It was after vespers one evening
When the vicar, inflamed by desire,
Beckoned a lad to the vestry,
Dismissing the rest of the choir.

He said, 'I've got something to show you,'
The boy followed hard on his heels,
Behind the locked door there was silence,
Except for some half-muffled squeals.

The vicar got two years (suspended),
The judge spoke of 'moral decay',
The vicar is sadder and wiser,
But the choir-boy is happy and gay.

413

Paradise Lost *as a Haiku*

Don't touch that fruit, Eve.
Oh my God—she's disobeyed!
Cosmic disaster!

ANONYMOUS

414

Limericks

(i)

There were three little owls in a wood,
Who sang hymns whenever they could.
  What the words were about,
  One could never make out,
But one felt it was doing them good.

(ii)

WHEN Daddy and Mum got quite plastered,
And their shame had been thoroughly mastered,
They told their boy, Harry:
'Son, we never *did* marry.
But don't tell the neighbours, you bastard.'

(iii)

THERE was aince an auld body o' Sydney
Wha suffered frae pains in the kidney.
He prayed tae the Lord
That he micht be restored,
And He promised He would—but He didnae!

(iv)

A YOUNG schizophrenic named Struther,
When told of the death of his brother,
Said: 'Yes, it's too bad,
But I can't feel too sad—
After all, I still have each other.'

(v)

THERE was an Archdeacon who said:
'May I take off my gaiters in bed?'
But the Bishop said: 'No,
Wherever you go
You must wear them until you are dead.'

PAUL DEHN

1912–1976

415 *Alternative Endings to an Unwritten Ballad*

I STOLE through the dungeons, while everyone slept,
Till I came to the cage where the Monster was kept.
There, locked in the arms of a Giant Baboon,
Rigid and smiling, lay . . . MRS RAVOON!

I climbed the clock-tower in the first morning sun
  And 'twas midday at least ere my journey was done;
But the clock never sounded the last stroke of noon,
  For there, from the clapper, swung MRS RAVOON.

I hauled in the line, and I took my first look
  At the half-eaten horror that hung from the hook.
I had dragged from the depths of the limpid lagoon
  The luminous body of MRS RAVOON.

I fled in the storm, thorough lightning and thunder,
  And there, as a flash split the darkness asunder,
Chewing a rat's-tail and mumbling a rune,
  Mad in the moat squatted MRS RAVOON.

I stood by the waters so green and so thick,
  And I stirred at the scum with my old, withered stick;
When there rose through the ooze, like a monstrous balloon,
  The bloated cadaver of MRS RAVOON.

Facing the fens, I looked back from the shore
  Where all had been empty a moment before;
And there, by the light of the Lincolnshire moon,
  Immense on the marshes, stood . . . MRS RAVOON!

416 from *Potted Swan*

(A condensed version of Shakespeare)

THE devil damn thee black, thou cream-faced loon—
Whom we invite to see us crowned at Scone.

ANONYMOUS

417 *Examination Question*

O CUCKOO! shall I call thee Bird
Or but a wandering Voice?
State the alternative preferred
With reasons for your choice.

ROY FULLER

1912–1991

418 *Coptic Socks*

The Victoria and Albert Museum has mounted a special exhibition of knitting ... The exhibits will include a pair of Coptic socks dating from the fourth–fifth century AD . . .

—NEWS ITEM, 1980

FANCY the Copt
Possessing socks!
—Elastic-topped,
Perhaps, with clocks.

What marvellous wool
From Coptic flocks
To last so well
In Coptic socks!

Some will get shocks
Who cast an optic
On knitted socks—
Then read they're Coptic.

VIRGINIA GRAHAM

1912–1992

419 *Ein Complaint*

HERR DIREKTOR, ich sent Sie ein cable,
zu frag' if some Dinge of mein
in ein drawer in der Ankleide table
gefunden sind. Ja oder nein?

Ein wunderschön Paar Underpanten;
ein Nachtgown von Satin gemacht;
und zwei kleine jade Elephanten,
mein Mann von der Ost mir gebracht.

Und auch, Herr Direktor, ich dinks
  dass ein Taschentuch shpotted mit green
war left in ein Stuhl on die links
  of die Lounge bei das photo of Wien.

Ich cabled der Tag before gestern:
  Warum kommst es nothing dabei?
Ich bitte Sie machen Ihr bestern
  zu schicken mir etwas reply!

LAWRENCE DURRELL

1912–1990

420 *Ballad of the Oedipus Complex*

FROM Travancore to Tripoli
I trailed the great Imago,
Wherever Freud has followed me
I felt Mama and Pa go.

(The engine loves the driver
And the driver loves his mate,
The mattress strokes the pillow
And the pencil pokes the slate.)

I tried to strangle it one day
While sitting in the Lido
But it got up and tickled me
And now I'm all Libido.

My friends spoke to the Censor
And the censor warned the Id
But though they tried to hush things up
They neither of them did.

(The barman loves his potion
And the admiral his barge,
The frogman loves the ocean
And the soldier his discharge.)

(The critic loves urbanity
The plumber loves his tool.
The preacher all humanity
The poet loves the fool.)

If seven psychoanalysts
On seven different days
Condemned my coloured garters
Or my neo-Grecian stays,

I'd catch a magic constable
And lock him behind bars
To be a warning to all men
Who have mamas and pas.

MICHAEL BURN

1912–

421 *For the Common Market*

IT'S easy to be witty in French.
You don't have to know French well.
Think of those French expressions (this is the secret) . . .
Goût du néant, esprit de l'escalier,
Dégoût de la vie, nostalgie de la boue,
Adieu suprême des mouchoirs.
All you have to do is take two nouns,
Any old nouns, the iller-assorted the better,
And couple them with a genitive,
Shrug, throw your hands out (not too far),
In a French sort of way,
And give the casual knowing look of someone
Who knows the girl at the bar.
Try it and see . . .
With faint disdain . . . c'est un sentiment de vestiaire,
Amour de boulanger, fantaisie du lavabo,
Goût de Londres, tendresse des wagons-lits,
Or sighing

Les aurevoirs de Vendredi.
Everyone will say how well you know French.
I've tried it on Frenchmen, and I know.

In German just couple the words together,
Like any old strangers meeting in any old street . . .
Himmelschnabel·Apfelpudel Heldenbegeisterung Weltkrebs . . .
No one will know any better.

In Italian it will be helpful to know the first line of Dante,
And also, brushing away a tear,
Italia, Italia, terra di morti,
And go straight on to business.

I will advise later about the Scandinavian countries.

DYLAN THOMAS

1914–1953

422 *The Song of the Mischievous Dog*

(Written at the age of 11)

THERE are many who say that a dog has his day,
And a cat has a number of lives;
There are others who think that a lobster is pink,
And that bees never work in their hives.
There are fewer, of course, who insist that a horse
Has a horn and two humps on its head,
And a fellow who jests that a mare can build nests
Is as rare as a donkey that's red.
Yet in spite of all this, I have moments of bliss,
For I cherish a passion for bones,
And though doubtful of biscuit, I'm willing to risk it,
And love to chase rabbits and stones.
But my greatest delight is to take a good bite
At a calf that is plump and delicious;
And if I indulge in a bite at a bulge,
Let's hope you won't think me too vicious.

JOHN BERRYMAN

1914–1972

423 *American Lights, Seen from off Abroad*

BLUE go up & blue go down
to light the lights of Dollartown

Nebuchadnezzar had it so good?
wink the lights of Hollywood

I never think, I have so many things,
flash the lights of Palm Springs

I worry like a madwoman over all the world,
affirm the lights, all night, at State

I have no plans, I mean well,
swear the lights of Georgetown

I have the blind staggers
call the lights of Niagara

We shall die in a palace
shout the black lights of Dallas

I couldn't dare less, my favorite son,
fritter the lights of Washington

(I have a brave old So-and-So,
chuckle the lights of Independence, Mo.)

I cast a shadow, what I mean,
blurt the lights of Abilene

Both his sides are all the same
glows his grin with all but shame

He can do nothing night & day,
wonder his lovers. So they say.

'Basketball in outer space'
sneers the White New Hampshire House

I'll have a smaller one, later, Mac,
hope the strange lights of Cal Tech

I love you one & all, hate shock,
bleat the lights of Little Rock

I cannot quite focus
cry the lights of Las Vegas

I am a maid of shots & pills,
swivel the lights of Beverly Hills

Proud & odd, you give me vertigo,
fly the lights of San Francisco

I am all satisfied love & chalk,
mutter the great lights of New York

I have lost your way
say the white lights of Boston

Here comes a scandal to blight you to bed.
Here comes a cropper. That's what I said.

R. P. LISTER

1914–

424

A Toast to 2,000

THE century's no longer new;
The years to come seem very few.
Twenties and thirties, forties gone,
And now the fifties rumble on . . .
No use to grumble or repine,
The century's in its decline.

Now dawns upon the turning page
The fin-de-siècle, stuffy age.

Young men and maidens of this time
Will be the pillars of its prime;
These jocund children, bald and stout,
Will see its last convulsions out.

And we who saw the thirties through,
The hungry forties suffered too,
May linger, grey and comatose,
Within a few years of its close;
But not behold the strange new years
Charged with fresh follies and fresh fears.

Yet some Victorian, shrunk and thin,
Will see the year 2000 in—
With fumbling mind, but changeless mien
Will ponder on the dear old Queen,
Under whose reign he first beheld
The frightening world, and wisely yelled.

There will he sit like any ghost
And drink to that New Year a toast,
Toast given by some pompous bore
At present playing on the shore.
Well may that centenarian fail
To grasp the meaning of the tale.

425 *A Mind Reborn in Streatham Common*

I RODE to Streatham Common on a tram,
  When I was young—how long ago it seems!—
And told a stout psychiatrist called Sam
  The sorry tale of my distressful dreams.

He with strong counsel straightened out my mind,
  And with shrewd blows unbent my twisted brain;
My eye grew placid and my brow unlined,
  And in that happy state I still remain.

My mind, unblemished by the slightest kink,
  Is like some plateau, featureless and bare;
I dream no more, I lie awake and think,
  But what I think is neither here nor there.

HENRY REED

1914–1986

426 *Chard Whitlow*

(*Mr Eliot's Sunday Evening Postscript*)

As we get older we do not get any younger.
Seasons return, and today I am fifty-five,
And this time last year I was fifty-four,
And this time next year I shall be sixty-two.
And I cannot say I should care (to speak for myself)
To see my time over again—if you can call it time,
Fidgeting uneasily under a draughty stair,
Or counting sleepless nights in the crowded Tube.

There are certain precautions—though none of them very reliable—
Against the blast from bombs, or the flying splinter,
But not against the blast from Heaven, *vento dei venti*,
The wind within a wind, unable to speak for wind;
And the frigid burnings of purgatory will not be touched
By any emollient.
I think you will find this put,
Better than I could ever hope to express it,
In the words of Kharma: 'It is, we believe,
Idle to hope that the simple stirrup-pump
Can extinguish hell.'
Oh, listeners,
And you especially who have turned off the wireless,
And sit in Stoke or Basingstoke, listening appreciatively to the silence,
(Which is also the silence of hell) pray, not for yourselves but your souls.
And pray for me also under the draughty stair.
As we get older we do not get any younger.

And pray for Kharma under the holy mountain.

HARRY HEARSON

fl. 1940

427 *Nomenclaturik*

THERE was a young fellow named Cholmondeley,
Whose bride was so mellow and colmondeley
That the best man, Colquhoun,
An inane young bolqufoun,
Could only stand still and stare dolmondeley.

The bridgeroom's first cousin, young Belvoir,
Whose dad was a Lancashire welvoir,
Arrived with George Bohun
At just about nohun
When excitement was mounting to felvoir.

The vicar—his surname was Beauchamp—
Of marriage endeavoured to teauchamp,
While the bridesmaid, Miss Marjoribanks,
Played one or two harjoripranks;
But the shoe that she threw failed to reauchamp.

GAVIN EWART

1916–

428 *The Black Box*

As well as these poor poems
I am writing some wonderful ones.
They are all being filed separately,
nobody sees them.

When I die they will be buried
in a big black tin box.
In fifty years' time
they must be dug up,

for so my will provides.
This is to confound the critics
and teach everybody
a valuable lesson.

429 *To the Virgins, to Make the Most of Time*

Now, listen.
I want you new girls, every morning,
To sprinkle an oral contraceptive on your corn flakes.
I've got my eye on you, I want to marry into you,
To fluffle you up a bit, then dive right in
Smoothly.

I'm a potentate. Don't be too girlish,
Don't bother to name those breasts Maria and Matilda
Or call your favourite ball-point Clarence.
None of this interests me. Wear a bra if you want to
And panties if you want to. It's immaterial.

In this establishment, my will holds.
If you are naughty, there's a cane in the corner.
I don't believe in God, I can do what I like.

Every morning there's naked bathing
And then at least two hours of horse-back riding
To promote a well-developed, rounded bottom.

The only lessons are Theory and Practice;
All my instructresses are big harsh Lesbians.
So watch your step.

Night Duty begins at eight. A roster will be published.
My favourite girls have a really marvellous time.
I hope you will be happy here. Never forget
These are the best years of your life.
Go to your rooms now. Goodnight.

430 *One for the Anthologies*

HERBERT's a hard and horrid man
And so am I.
He does as much harm as he can
And so do I.
He wastes the time of Institutes
And spends his nights with prostitutes,
And so do I.

Wilfred's a weak and weary man
And so am I.
He's always been an also-ran
And so have I.
He's been defeated all his life,
Too tired to end it with a knife—
And so am I.

David's a dense and drunken man
And so am I.
He's fond of glass and mug and can
And so am I.
When these sad dogs have had their day
They'll all be glad to go away
And so will I.

431 *The Great Women Composers*

SYBIL SIBELIUS! Yes, Belinda Brahms!
Harriet Haydn! And even Mary Mozart!
Have all been invited to tea
by the indomitable Beatrice K. Beethoven!

432 *The Semantic Limerick According to the Shorter Oxford English Dictionary (1933)*

THERE existed an adult male person who had lived a relatively short time, belonging or pertaining to St John's,* who desired to commit sodomy with the large web-footed swimming birds of the genus *Cygnus* or subfamily *Cygninae* of the family *Anatidae*, characterized by a long and gracefully curved neck and a majestic motion when swimming.

So he moved into the presence of the person employed to carry burdens, who declared: 'Hold or possess as something at your disposal my female child! The large web-footed swimming-birds of the genus *Cygnus* or subfamily *Cygninae* of the family *Anatidae*, characterized by a long and gracefully curved neck and a majestic motion when swimming, are set apart, specially retained for the Head, Fellows and Tutors of the College!'

*A College of Cambridge University.

433 *The Semantic Limerick According to Dr Johnson's Dictionary (Edition of 1765)*

THERE exifted a person, not a woman or a boy, being in the firft part of life, not old, of St John's,* who wifhed to —— the large water-fowl, that have a long and very ftraight neck, and are very white, excepting when they are young (their legs and feet being black, as are their bills, which are like that of a goofe, but fomething rounder, and a little hooked at the lower ends, the two fides below their eyes being black and fhining like ebony).

In confequence of this he moved ftep by ftep to the one that had charge of the gate, who pronounced: 'Poffefs and enjoy my female offspring! The large water-fowl, that have a long and very ftraight neck, and are very white, excepting when they are young (their legs and feet being black, as are their bills, which are like that of a goofe, but fomething rounder, and a little hooked at the lower ends, the two fides below their eyes being black and fhining like ebony), are kept in ftore, laid up for a future time, for the fake of the gentlemen with Spanish titles.'

*A College of Cambridge University.

434 *The Owl Writes a Detective Story*

This poem was written to be read aloud, and the 'oo' sounds at the ends of the lines should be intoned like the call of an owl

A STATELY home where doves, in dovecotes, coo—
fields where calm cattle stand and gently moo,
trim lawns where croquet is the thing to do.
This is the ship, the house party's the crew:
Lord Feudal, hunter of the lion and gnu,
whose walls display the heads of not a few,
Her Ladyship, once Ida Fortescue,
who, like his Lordship very highborn too
surveys the world with a disdainful moue.
Their son—most active with a billiard cue—
Lord Lazy (stays in bed till half past two).
A Balkan Count called Popolesceru
(an ex-Dictator waiting for a coup).
Ann Fenn, most English, modest, straight and true,
a very pretty girl without a sou.
Adrian Finkelstein, a clever Jew.
Tempest Bellairs, a beauty such as you
would only find in books like this (she'd sue
if I displayed her to the public view—
enough to say men stick to her like glue).
John Huntingdon, who's only there to woo
(a fact, except for her, the whole house knew)
Ann Fenn. And, last, the witty Cambridge Blue,
the Honourable Algy Playfair, who
shines in detection. His clear 'View halloo!'
puts murderers into a frightful stew.

But now the plot unfolds! What *déjà vu*!
There! In the snow!—The clear print of a shoe!
Tempest is late for her next rendez-vous,
Lord Feudal's blood spreads wide—red, sticky goo
on stiff white shirtfront—Lazy's billet-doux
has missed Ann Fenn, and Popolesceru
has left—without a whisper of adieu
or saying goodbye, typical *mauvais gout*!
Adrian Finkelstein, give him his due,
behaves quite well. Excitement is taboo
in this emotionless landowner's zoo.

Algy, with calm that one could misconstrue
(handling with nonchalance bits of vertu)
knows who the murderer is. He has a clue.

But who? But who? Who, who, who, who, who, who?

435

'*It's Hard to Dislike Ewart*'

—*New Review* critic

I ALWAYS try to dislike my poets,
it's good for them, they get so uppity otherwise,
going around thinking they're little geniuses—
but sometimes I find it hard. They're so pathetic
in their efforts to be *liked*.

When we're all out walking on the cliffs
it's always pulling my coat with 'Sir! Oh, Sir!'
and 'May I walk with *you*, Sir?'—
I sort them out harshly with my stick.

If I push a few over the edge, that only
encourages the others. In the places of preferment
there is room for just so many.
The rest must simply lump it.
There's too much sucking up and trying to be clever.
They must all learn they'll never get round *me*.
Merit has nothing to do with it. There's no way
to pull the wool over my eyes, *no* way,
no way . . .

ROBERT CONQUEST

1917–

436

Bagpipes at the Biltmore

(*Air*: 'Strathfiddich')

DOWNTOWN Los Angeles:
In the huge baroque lobby
Like the hall of a station
Three men with kilts on,

Skian-dhus and sporrans
And all the adornments,
Stamp round in circles.
One bangs a drum and
The others play bagpipes.

What on earth are they up to?
And who are the people
Marching behind them
With badges and name-tags?
An occasional banner:
'UBEW'—
'United', for certain,
'Brotherhood', surely,
And W's 'Workers'.
What about 'E', then?
'Electrical', maybe?
But what has the Union
To do with bagpipes?
And why are they here in
The bourgeois old Biltmore?

It's terribly noisy
But fairly inspiriting
—Except, I remember
How we marched as recruits through
The fog-frozen Lothians
With pipers before us:
An ear-splitting torment
Till two brave English laddies
Put paid to the nuisance
With a long knitting-needle,
And the fights and the uproar
Around the old squad-room ...

My highland great-grandsires
—Macandrew, Macpherson—
Though they much preferred Mozart,
Passed down an old story
Of the head of a septlet
In a bare granite hovel
He miscalled a castle.
After plenty of skirlings
His hereditary piper
With a tattered old plaid on

Would cry to the cardinal
Points of the compass,
'The MacShagbag of Shagbag
Has started to dine.
The Kings of the Airth
May now take their seats'.

—For that you need bagpipes.
Mere whisky won't do it.
So they do have their virtues,
At least of nostalgia.
But instilling a spirit
In Electrical Workers
To march round the lobby
Of the stuffy old Biltmore
I believe it, since I see it …

Just one more blow from
The unreasoned, untidy
World we inhabit
Against our assumptions.
Good for you, really:
Shakes up your smugness,
Freshens your senses
(In this case your eardrums).
I expect that tomorrow
Will turn up with something,
Though perhaps a bit subtler
Than this too sharp reminder
To haul off from habit
And struggle with structure,
With a noise so enormous
I can bellow unnoticed
(And it's pleasant to do it
In the solemn old Biltmore),
'Oh belt up, you buggers!'

437

Progress

THERE was a great Marxist called Lenin
Who did two or three million men in
—That's a lot to have done in
But where he did one in
That grand Marxist Stalin did ten in.

JOHN HEATH-STUBBS

1918–

438 *Simcox*

SIMCOX was one of several rather uninteresting
Ghosts, which popular report affirmed
Haunted the precincts of the College where
I had the privilege of my education.
A Junior Fellow (exactly in what field
His tedious studies ran was not remembered),
Simcox, it seems, was drowned—the Long Vacation
Of 1910, or '12, or thereabouts,
Somewhere off the coast of Donegal:
If accident or suicide I don't recall. But afterwards
Simcox began to manifest himself
In his old rooms, sitting in his large arm-chair,
With dripping clothes, and coughing slightly.
Simcox was wet in life, and wet in death.

To save embarrassment, it was decided
This should in future be the chaplain's room.
The chaplain of my day, a hearty
Beer-swilling extrovert, and not much given—
Or so I would suppose—to exorcism,
Never, to my knowledge, did in fact aver
He had encountered Simcox. And anyway
Those visitings grew fainter with the years.
Simcox was dim in life, and dim in death.

439 *One*

ONE thinks of *one* as a pronoun employed principally
At Cambridge, modestly to include oneself
And other people in one's own set,
At Cambridge. One appreciates the French usage
Of *on*; one knows one's Henry James;
One does feel (or, of course, alternatively, one does not)
One must, on the whole, concur with Dr Leavis
(or, of course, alternatively, with Mr Rylands).

At Oxford, on the other hand,
One tends to become *we*. At Cambridge
One senses a certain arrogance in the Oxford *we*;
A certain exclusiveness in the Cambridge *one*
Is suspected, at Oxford.

440 *Footnote to Belloc's 'Tarantella'*

Do you remember an inn, Miranda? It lost its licence of course—
Total neglect of the elementary rules of hygiene, not to mention
Wine contaminated with tar, and constant complaints
From the neighbours about the noise.
One visit of the inspector
From the Spanish Tourist Board was quite enough.
Nevermore, Miranda, nevermore.

MARTIN BELL

1918–1978

441 *Senilio Passes, Singing*

SOLOMON GRUNDY
Bored on Tuesday
Manic on Wednesday
Panic on Thursday
Drunk on Friday
Hung over Saturday
Slept all Sunday
Back to work Monday—
That's the life
For Solomon Grundy.

LOUISE BENNETT

1919–

442

Colonization in Reverse

WAT a joyful news, Miss Mattie,
I feel like me heart gwine burs
Jamaica people colonizin
Englan in reverse.

By de hundred, by de tousan
From country and from town,
By de ship-load, by de plane-load
Jamaica is Englan boun.

Dem a pour out a Jamaica,
Everybody future plan
Is fe get a big-time job
An settle in de mother lan.

What a islan! What a people!
Man an woman, old an young
Just a pack dem bag an baggage
An tun history upside dung!

Some people doan like travel,
But fe show dem loyalty
Dem all a open up cheap-fare-
To-Englan agency.

An week by week dem shippin off
Dem countryman like fire,
Fe immigrate an populate
De seat a de Empire.

Oonoo see how life is funny,
Oonoo see de tunabout?
Jamaica live fe box bred
Out a English people mout'.

oonoo] you

For wen dem ketch a Englan,
An start play dem different role,
Some will settle down to work
An some will settle fe de dole.

Jane say de dole is not too bad
Because dey payin she
Two pounds a week fe seek a job
Dat suit her dignity.

Me say Jane will never fine work
At de rate how she dah look,
For all day she stay pon Aunt Fan couch
An read love-story book.

Wat a devilment a Englan!
Dem face war an brave de worse,
But me wonderin how dem gwine stan
Colonizin in reverse.

WILLIAM COLE

1919–

443

Mutual Problem

SAID Jerome K. Jerome to Ford Madox Ford,
'There's something, old boy, that I've always abhorred:
When people address me and call me "Jerome,"
Are they being standoffish, or too much at home?'

Said Ford, 'I agree;
It's the same thing with me.'

ketch] reach

ANONYMOUS

444 *Harry Pollitt Was a Bolshie*

HARRY POLLITT was a Bolshie, one of Lenin's lads,
Till he was finally done in by reactionary cads.

By reactionary cads? Yes, reactionary cads,
Till he was finally done in by reactionary cads.

Up spoke the ghost of Harry, 'My spirit shall not die,
I'll go and do some party work in the kingdom up on high.'

He stood before the pearly gates, all trembling at the knees,
'A message here for Comrade God from Harry Pollitt, please.'

'Oh, who is that a-standing there so humble and contrite?'
'I'm a friend of Lady Astor's.' 'Okay, you'll be all right.'

They put him in the choir, put a harp into his hand,
And he taught the Internationale to the Hallelujah band.

They put him in the choir, but the hymns he did not like,
So he organised the angels and he brought them out on strike.

They sent him up for trial before the Holy Ghost,
For spreading disaffection among the Heavenly Host.

The verdict it was 'Guilty,' and Harry he said 'Oh!'
He tucked his nightie round his knees and he floated down below.

Yes, he floated down below, he floated down below,
He tucked his nightie round his knees and he floated down below.

Now seven long years have passed, and Harry's doing well—
He's just been made the People's Commissar of Soviet Hell.

Harry Pollitt] General Secretary of the Communist Party of Great Britain 1929–56

ALEXANDER SCOTT

1920–

445

from *Scotched*

Scotch God
Kent His
Faither.

Scotch Religion
Damn
Aa.

Scotch Education
I tellt ye
I tellt ye.

Scotch Queers
Wha peys wha
—For what?

Scotch Prostitution
Dear,
Dear.

Scotch Liberty
Agree
Wi me.

Scotch Equality
Kaa the feet frae
Thon big bastard.

Scotch Fraternity
Our mob uses
The same razor.

Scotch Optimism
Through a gless,
Darkly.

Kent] knew *Kaa*] knock *Thon*] that

Scotch Pessimism
Nae
Gless.

Scotch Initiative
Eftir
You.

Scotch Generosity
Eftir
Me.

Scotch Sex
In atween
Drinks.

Scotch Passion
Forgot
Mysel.

Scotch Free-Love
Canna be
Worth much.

Scotch Lovebirds
Cheap
Cheap.

Scotch Fractions
A hauf
'n' a hauf.

Scotch Afternoon-Tea
Masked
Pot

Scotch Drink
Nip
Trip.

Scotch Poets
Wha's the
T'ither?

Masked] infused

HOWARD NEMEROV

1920–1991

446

A Life

INNOCENCE?
In a sense.
In no sense!

Was that *it*?
Was *that* it?
Was that it?

That was it.

447

Creation Myth on a Moebius Band

THIS world's just mad enough to have been made
By the Being his beings into Being prayed.

448

To my Least Favorite Reviewer

WHEN I was young, just starting at our game,
I ambitioned to be christlike, and forgive thee.
For a mortal Jew that proved too proud an aim;
Now it's my humbler hope just to outlive thee.

EDWIN MORGAN

1920–

449

Itinerary

I

WE went to Oldshoremore.
Is the Oldshoremore road still there?
You mean the old shore road?
I suppose it's more an old road than a shore road.

No more! They shored it up, but it's washed away.
So you could sing the old song—
Yes we sang the old song:
  We'll take the old Oldshoremore shore road no more.

II

We passed the Muckle Flugga.
Did you see the muckle flag?
All we saw was the muckle fog.
The flag says ULTIMA FLUGGA WHA'S LIKE US.
Couldn't see flag for fug, sorry.
Ultimately—
  Ultimately we made for Muck and flogged the lugger.

III

Was it bleak at Bowhousebog?
It was black as a hoghouse, boy.
Yes, but bleak?
Look, it was black as a bog and bleak as the Bauhaus!
The Bauhaus wasn't black—
Will you get off my back!
So there were dogs too?
  Dogs, hogs, leaks in the bogs—we never went back.

D. J. ENRIGHT

1920–

450

Posterity

POSTERITY was always a great reader.
He would beg, borrow or steal books,
He would even buy them.
You could be sure to find Posterity
With his nose in a book.
(Except when listening to music
Or peering at paintings.)
He had excellent judgement too.
You could always put your faith in old Posterity.
We shall miss him.

451 from *Paradise Illustrated*

(i)

'RICH soil,' remarked the Landlord.
'Lavishly watered.' Streams to the right,
Fountains to the left.
'The rose, you observe, is without a thorn.'

'What's a thorn?' asked Adam.
'Something you have in your side,'
The Landlord replied.

'And since there are no seasons
All the flowers bloom all the time.'

'What's a season?' Eve inquired.
'Yours not to reason why,'
The Landlord made reply.

Odours rose from the trees,
Grapes fell from the vines,
The sand was made of gold,
The pebbles were made of pearls.

'I've never seen the like,' said Eve.
'Naturally,' the Landlord smiled.

'It's unimaginable!' sighed Adam.
'You're not obliged to imagine it,'
Snapped the Landlord. 'Yet.'

(ii)

'WHY didn't we think of clothes before?'
Asked Adam,
Removing Eve's.

'Why did we ever think of clothes?'
Asked Eve,
Laundering Adam's.

(iii)

THE days of Adam were 930 years.
He sat in the market-place
With other senior citizens.

With Seth, who was just 800,
Enos, who was 695,
And Methuselah, only 243.

'They're not the men
Their great-grandfathers were,'
Said Seth.

'Lamech's kid Noah cries all night,'
Said Enos.
'Howls when they bath it.'

Said the youthful Methuselah:
'They've all been spoilt.
I blame their mothers.'

'It was different in my day,'
Said Adam.
'People lived for ever then.'

452 *The Evil Days*

WHILE the years draw nigh when the clattering typewriter is a burden, likewise a parcel of books from the postman, and he shall say, I have no pleasure in them; for much study is a weariness of the flesh;

Also when the cistern shall break, and the overflow be loosed like a fountain; when the lights are darkened, and the windows need cleaning,

And the keeper of the house shall tremble at rate bills, and be afraid of prices which are high; and almonds are too much for the grinders, and beers shall be out of the way;

Yet desire may not utterly fail, and he shall rise up at the sight of a bird, when the singsong girls are brought low;

In the day when he seeks out acceptable words; when the editors are broken at their desks, and the sound of the publishers shall cease because they are few.

Then shall the dust return to the earth as it was, and the spirit also, whether it be good or evil, shall look for its place.

RICHARD WILBUR

1921–

453 *Rillons, Rillettes*

RILLETTES: Hors d'œuvre *made up of a mash of pigmeat, usually highly seasoned. Also used for making sandwiches. The* Rillettes *enjoying the greatest popularity are the* Rillettes *and* Rillons de Tours, *but there are* Rillettes *made in many other parts of France.*

RILLONS: *Another name for the* Rillettes, *a pigmeat* hors d'œuvre. *The most popular* Rillons *are those of Blois.*

—A Concise Encyclopaedia of Gastronomy, *edited by*
André L. Simon

RILLONS, *Rillettes*, they taste the same,
And would by any other name,
And are, if I may risk a joke,
Alike as two pigs in a poke.

The dishes are the same, and yet
While Tours provides the best *Rillettes*,
The best *Rillons* are made in Blois.
There must be some solution.
Ah!—

Does Blois supply, do you suppose,
The best *Rillettes de Tours*, while those
Now offered by the chefs of Tours
Are, by their ancient standards, poor?

Clever, but there remains a doubt.
It is a thing to brood about,
Like non-non-A, infinity,
Or the doctrine of the Trinity.

454 *The Prisoner of Zenda*

AT the end a
'The Prisoner of Zenda,'
The King being out of danger,
Stewart Granger
(As Rudolph Rassendyll)

Must swallow a bitter pill
By renouncing his co-star,
Deborah Kerr.

It would be poor behavia
In him and in Princess Flavia
Were they to put their own
Concerns before those of the Throne.
Deborah Kerr must wed
The King instead.

Rassendyll turns to go.
Must it be so?
Why can't they have their cake
And eat it, for heaven's sake?
Please let them have it both ways,
The audience prays.
And yet it is hard to quarrel
With a plot so moral.

One redeeming factor,
However, is that the actor
Who plays the once-dissolute King
(Who has learned through suffering
Not to drink or be mean
To his future Queen),
Far from being a stranger,
Is *also* Stewart Granger.

455 *Shame*

IT is a cramped little state with no foreign policy,
Save to be thought inoffensive. The grammar of the language
Has never been fathomed, owing to the national habit
Of allowing each sentence to trail off in confusion.
Those who have visited Scusi, the capital city,
Report that the railway-route from Schuldig passes
Through country best described as unrelieved.
Sheep are the national product. The faint inscription
Over the city gates may perhaps be rendered,
'I'm afraid you won't find much of interest here.'

Census-reports which give the population
As zero are, of course, not to be trusted,
Save as reflecting the natives' flustered insistence
That they do not count, as well as their modest horror
Of letting one's sex be known in so many words.
The uniform grey of the nondescript buildings, the absence
Of churches or comfort-stations, have given observers
An odd impression of ostentatious meanness,
And it must be said of the citizens (muttering by
In their ratty sheepskins, shying at cracks in the sidewalk)
That they lack the peace of mind of the truly humble.
The tenor of life is careful, even in the stiff
Unsmiling carelessness of the border-guards
And *douaniers*, who admit, whenever they can,
Not merely the usual carloads of deodorant
But gypsies, g-strings, hasheesh, and contraband pigments.
Their complete negligence is reserved, however,
For the hoped-for invasion, at which time the happy people
(Sniggering, ruddily naked, and shamelessly drunk)
Will stun the foe by their overwhelming submission,
Corrupt the generals, infiltrate the staff,
Usurp the throne, proclaim themselves to be sun-gods,
And bring about the collapse of the whole empire.

PHILIP LARKIN

1922–1985

456

Sunny Prestatyn

Come To Sunny Prestatyn
Laughed the girl on the poster,
Kneeling up on the sand
In tautened white satin.
Behind her, a hunk of coast, a
Hotel with palms
Seemed to expand from her thighs and
Spread breast-lifting arms.

She was slapped up one day in March.
A couple of weeks, and her face
Was snaggle-toothed and boss-eyed;
Huge tits and a fissured crotch
Were scored well in, and the space
Between her legs held scrawls
That set her fairly astride
A tuberous cock and balls

Autographed *Titch Thomas*, while
Someone had used a knife
Or something to stab right through
The moustached lips of her smile.
She was too good for this life.
Very soon, a great transverse tear
Left only a hand and some blue.
Now *Fight Cancer* is there.

457

A Study of Reading Habits

WHEN getting my nose in a book
Cured most things short of school,
It was worth ruining my eyes
To know I could still keep cool,
And deal out the old right hook
To dirty dogs twice my size.

Later, with inch-thick specs,
Evil was just my lark:
Me and my cloak and fangs
Had ripping times in the dark.
The women I clubbed with sex!
I broke them up like meringues.

Don't read much now: the dude
Who lets the girl down before
The hero arrives, the chap
Who's yellow and keeps the store,
Seem far too familiar. Get stewed:
Books are a load of crap.

458

Limericks

(i)

THERE was an old fellow of Kaber,
Who published a volume with Faber:
  When they said, 'Meet Ted Hughes,'
  He replied, 'I refuse,'
But Charles called, 'You must love your neighbour.'

(ii)

THERE was an old fellow of Kaber,
Who published a volume with Faber:
  When they said, 'Meet Thom Gunn,'
  He cried, 'God, I must run,'
But Charles called, 'You must love your neighbour.'

Larkin wrote a number of other limericks in the same vein.

HOWARD MOSS

1922–1987

459

Geography: A Song

THERE are no rocks
At Rockaway,
There are no sheep
At Sheepshead Bay,
There's nothing new
In Newfoundland,
And silent is
Long Island Sound.

460

Tourists

CRAMPED like sardines on the Queens, and sedated,
The sittings all first, the roommates mismated,

Kaber] a village in Yorkshire   *Charles*] Charles Monteith, editor at Faber and Faber.

Three nuns at the table, the waiter a barber,
Then dumped with their luggage at some frumpish harbor,

Veering through rapids in a vapid *rapido*
To view the new moon from a ruin on the Lido,

Or a sundown in London from a rundown Mercedes,
Then high-borne to Glyndebourne for Orfeo in Hades,

Embarrassed in Paris in Harris tweed, dying to
Get to the next museum piece that they're flying to,

Finding, in Frankfurt, that one indigestible
Comestible makes them too ill for the Festival,

Footloose in Lucerne, or taking a pub in in
Stratford or Glasgow, or maudlin in Dublin, in-

sensitive, garrulous, querulous, audible,
Drunk in the Dolomites, tuning a portable,

Homesick in Stockholm, or dressed to toboggan
At the wrong time of year in too dear Copenhagen,

Generally being too genial or hostile—
Too grand at the Grand, too old at the Hostel—

Humdrum conundrums, what's to become of them?
Most will come home, but there will be some of them

Subsiding like Lawrence in Florence, or crazily
Ending up tending shop up in Fiesole.

KINGSLEY AMIS

1922–

461

Autobiographical Fragment

WHEN I lived down in Devonshire
  The callers at my cottage
Were Constant Angst, the art critic,
  And old Major Courage.

Angst always brought me something nice
To get in my good graces:
A quilt, a roll of cotton-wool,
A pair of dark glasses.

He tore up all my unpaid bills,
Went and got my slippers,
Took the telephone off its hook
And bolted up the shutters.

We smoked and chatted by the fire,
Sometimes just nodding;
His charming presence made it right
To sit and do nothing.

But then—those awful afternoons
I walked out with the Major!
I ran up hills, down streams, through briars;
It was sheer blue murder.

Trim in his boots, riding-breeches
And threadbare Norfolk jacket,
He watched me, frowning, bawled commands
To work hard and enjoy it.

I asked him once why I was there,
Except to get all dirty;
He tugged his grey moustache and snapped:
'Young man, it's your duty.'

What duty's served by pointless, mad
Climbing and crawling?
I tell you, I was thankful when
The old bore stopped calling.

462 *Mightier than the Pen*

JERKING and twitching as he walks,
Neighing and hooting as he talks,
The shabby pundit's prototype,
Smoking his horrible black pipe,
Balbus keeps making me feel ill.
I've heard that art's a kind of pill

To purge your feelings, so I'll try
And catch him in my camera's eye,
Transcribe him down to the last hair,
Ambered, though neither rich nor rare.

But will my interests be served
By having such a sod preserved?
Is art much better than a drug,
To cure the man but spare the bug?
And, gentle reader, why should you
Be led vicariously to spew?
Cameras just click, and a click's not
The sound of an effective shot;
Fussing with flash and tripod's fun,
But bang's the way to get things done.

VERNON SCANNELL

1922–

463 *Popular Mythologies*

NOT all these legends, I suppose,
Are total lies:
Thousands of Welshmen at a rugby match
Sometimes surprise
By singing more or less in tune;
Quite a lot
Of bullies may be cowards, though too many
Are not.
It would make sense to revise most saws
About the tall:
For instance I've found that the bigger they are
The harder *I* fall.
Everyone knows about the Scots
And their miserliness,
The way they repeatedly cry 'Hoots mon!'
Though I must confess
I have often been embarrassed by their largesse
And not one
In my hearing has ever said 'Hoots mon'.

Of course the Jews
Are very easy to pin down:
They amuse
With comic names like Izzy, Solly,
Benny, Moses,
And like the Scots they're very mean
And have big noses.
The trouble is you can't be sure
That people with small
Noses and names like Patrick, Sean,
Peter or Paul
Are always to be trusted in fiscal matters.
Not at all.
And what of the famous cockney wit?
You'd hear greater
Intelligence, humour and verbal flair
From an alligator
Than the average East End Londoner.
And suicides?
Those who talk about it never do it?
What of the brides
Of darkness, Sexton, Woolf and Plath?
And those others
Chatterton, Beddoes, Hemingway, Crane?
This band of brothers
All descanted on the mortal theme.
But why go on thus?
Easy, but waste of effort to fill
An omnibus.
Surprising though how people still
Swallow all this
Prejudiced stuff—not us, of course,
Oh no, not us.

464

Protest Poem

IT was a good word once, a little sparkler,
Simple, innocent even, like a hedgerow flower,
And irreplaceable. None of its family
Can properly take over: *merry* and *jolly*
Both carry too much weight; *jocund* and *blithe*
Were pensioned off when grandpa was alive.
Vivacious is a flirt; she's lived too long

With journalists and advertising men.
Spritely and *spry*, both have a nervous tic.
There is no satisfactory substitute.
It's down the drain and we are going to miss it.
No good advising me to go ahead
And use the word as ever. If I did
We know that someone's bound to smirk or snigger.
Of all the epithets why pick on this one?
Some deep self-mocking irony
Or blindfold stab into the lexicon?
All right. Then let's call heterosexuals *sad*,
Dainty for rapists, *shy* for busy flashers,
Numinous for necrophiles, *quaint* for stranglers;
The words and world are mad! I must protest
Although I know my cause is lost.
A good word once, and I'm disconsolate
And angered by this simple syllable's fate:
A small innocence gone, a little Fall.
I grieve the loss. I am not gay at all.

'LORD BEGINNER'
(Egbert Moore)

fl. 1950

465

Victory Calypso

CRICKET, lovely cricket,
At Lord's where I saw it;
Cricket, lovely cricket,
At Lord's where I saw it;
Yardley tried his best
But Goddard won the Test.
They gave the crowd plenty fun;
Second Test and West Indies won.

With those two little pals of mine
Ramadhin and Valentine.

The King was there well attired,
So they started with Rae and Stollmeyer;
Stolly was hitting balls around the boundary,
But Wardle stopped him at twenty.
Rae had confidence,
So he put up a strong defence;
He saw the King was waiting to see,
So he gave him a century.

With those two little pals of mine
Ramadhin and Valentine.

West Indies first innings total was three-twenty-six
Just as usual.
When Bedser bowled Christiani
The whole thing collapsed quite easily,
England then went on,
And made one-hundred-fifty-one;
West Indies then had two-twenty lead,
And Goddard said, 'That's nice indeed.'

With those two little pals of mine
Ramadhin and Valentine.

Yardley wasn't broken-hearted
When the second innings started;
Jenkins was like a target
Getting the first five into his basket.
But Gomez broke him down,
While Walcott licked them around;
He was not out for one-hundred and sixty-eight,
Leaving Yardley to contemplate.

The bowling was super-fine
Ramadhin and Valentine.

West Indies was feeling homely,
Their audience had them happy.
When Washbrook's century had ended,
West Indies' voices all blended.
Hats went in the air.
They jumped and shouted without fear;
So at Lord's was the scenery
Bound to go down in history.

After all was said and done,
Second Test and West Indies won!

ANTHONY BUTTS

fl. 1950

466

Massenet

MASSENET
Never wrote a Mass in A.
It'd have been just too bad,
If he had.

ALLAN M. LAING

fl. 1950

467

A Grace for Ice-Cream

FOR water-ices, cheap but good,
That find us in a thirsty mood;
For ices made of milk or cream
That slip down smoothly as a dream;
For cornets, sandwiches and pies
That make the gastric juices rise;
For ices bought in little shops
Or at the kerb from him who stops;
For chanting of the sweet refrain:
'Vanilla, strawberry or plain?'
  We thank Thee, Lord, who sendst with heat
  This cool deliciousness to eat.

JUSTIN RICHARDSON

fl. 1950

468

The Retort Perfect

A RARE, twice-in-a-lifetime form of sport
Is the finally annihilating retort.

The sort of thing I have in mind
Is devastating but never at all unkind
And—which should be abundantly made clear—
It must be audible *and* now *and* here,
Not muttered tomorrow morning, to ourselves, in the train—
We can all do that.
No: there must be that glorious dawn in the brain
Then out it comes, perfect-born, pat
Clear-spoken, deliberate, cool
And—this is the test—so remote from all human reply
That it couldn't occur to our opposite number to try.
He mustn't be made to look a fool
Or he won't go about and repeat it;
He should just grin and beat it,
Leaving us amazed at our inspiration,
Saying it over and over again for sweet confirmation,
And rehearsing the tactics of its infiltration
Into all future conversation.

MICHAEL FLANDERS

1922–1975

469 *Have Some Madeira, M'dear?*

SHE was young! She was pure! She was new! She was nice!
She was fair! She was sweet seventeen!
He was old! He was vile and no stranger to vice!
He was base! He was bad! He was mean!
He had slyly inveigled her up to his flat
To see his collection of stamps,
And he said as he hastened to put out the cat,
The wine, his cigar and the lamps,

'Have some MADEIRA m'dear!
You really have nothing to fear;
I'm not trying to tempt you, that wouldn't be right,
You shouldn't drink spirits at this time of night,
Have some MADEIRA, m'dear!
It's very much nicer than BEER;

I don't care for SHERRY, one cannot drink STOUT,
And PORT is a wine I can well do without,
It's simply a case of "Chacun à son GOUT."
Have some MADEIRA, m'dear.'

Unaware of the wiles of the snake in the grass,
Of the fate of the maiden who topes,
She lowered her standards by raising her glass,
Her courage, her eyes and his hopes!
She sipped it, she drank it, she drained it, she did,
He quietly refilled it again,
And he said as he secretly carved one more notch
On the butt of his gold-handled cane,

'Have some MADEIRA, m'dear!
I've got a small cask of it here,
And once it's been opened you know it won't keep;
Do finish it up, it will help you to sleep;
Have some MADEIRA, m'dear!
It's really an excellent year;
Now it if were GIN you'd be wrong to say yes,
The evil GIN does would be hard to assess
(Beside it's inclined to affect m'prowess!)
Have some MADEIRA, m'dear!'

Then there flashed through her mind what her mother had said,
With her antepenultimate breath:
'Oh my child, should you look on the wine when 'tis red,
Be prepared for a fate worse than death!'
She let go her glass with a shrill little cry,
Crash, tinkle! It fell to the floor.
When he asked 'What in heaven . . .?' she made no reply,
Up her mind and a dash for the door . . .

'HAVE SOME MADEIRA M'DEAR!'
Rang out down the hall loud and clear,
A tremulous cry that was filled with despair,
As she paused to take breath in the cool midnight air;
'HAVE SOME MADEIRA, M'DEAR!'
The words seemed to ring in her ear;
Until the next morning she woke up in bed,
With a smile on her lips and an ache in her head
And a beard in her ear-hole that tickled and said:
'Have some MADEIRA, m'dear!'

ANTHONY HECHT

1923–

470

Goliardic Song

IN classical environrons
  Deity misbehaves;
There nereids and sirens
  Bucket the whomping waves.
As tritons sound their conches
  With fat, distended cheeks,
Welded are buxom haunches
  To muscular physiques.

Out of that frothy pageant
  Venus Pandemos rose,
Great genetrix and regent
  Of human unrepose.
Not age nor custom cripples
  Her strenuous commands,
Imperative of nipples
  And tyrannous of glands.

We who have been her students,
  Matriculated clerks
In scholia of imprudence
  And vast, venereal Works,
Taken and passed our orals,
  Salute her classic poise:
Ur-Satirist of Morals
  And Mother of our Joys.

471

An Old Malediction

(Freely from Horace)

WHAT well-heeled knuckle-head, straight from the unisex
Hairstylist and bathed in *Russian Leather*,
Dallies with you these late summer days, Pyrrha,
In your expensive sublet? For whom do you
Slip into something simple by, say, Gucci?
The more fool he who has mapped out for himself
The saline latitudes of incontinent grief.

Dazzled though he be, poor dope, by the golden looks
Your locks fetched up out of a bottle of *Clairol*,
He will know that the wind changes, the smooth sailing
Is done for, when the breakers wallop him broadside,
When he's rudderless, dismasted, thoroughly swamped
In that mindless rip-tide that got the best of me
Once, when I ventured on your deeps, Piranha.

472

From the Grove Press

HIGGLEDY-PIGGLEDY
Ralph Waldo Emerson
Wroth at Bostonian,
Cowardly hints,

Wrote an unprintable
Epithalamion
Based on a volume of
Japanese prints.

LOUIS SIMPSON

1923–

473

Chocolates

ONCE some people were visiting Chekhov.
While they made remarks about his genius
the Master fidgeted. Finally
he said, 'Do you like chocolates?'

They were astonished, and silent.
He repeated the question,
whereupon one lady plucked up her courage
and murmured shyly, 'Yes.'

'Tell me,' he said, leaning forward,
light glinting from his spectacles,
'what kind? The light, sweet chocolate
or the dark, bitter kind?'

The conversation became general.
They spoke of cherry centers,
of almonds and Brazil nuts.
Losing their inhibitions
they interrupted one another.
For people may not know what they think
about politics in the Balkans,
or the vexed question of men and women,
but everyone has a definite opinion
about the flavor of shredded coconut.
Finally someone spoke of chocolates filled with liqueur,
and everyone, even the author of *Uncle Vanya*,
was at a loss for words.

As they were leaving he stood by the door
and took their hands.
In the coach returning to Petersburg
they agreed that it had been a most
unusual conversation.

EDWARD FIELD

1924–

474 *Lower East Side: The George Bernstein Story*

It starts on the Lower East Side
when Irving Berlin, Fanny Brice, Paul Muni,
all the bigshots in show business, the underworld,
politics, and the arts,
were still rollerskating among the pushcarts,
and calling up Hey Ma for a penny to be thrown down
and if no penny fell, going off to earn one by their talents.

Just off the boat the Bernsteins arrive,
corner Delancey and Orchard, greenhorns,
Mama with her big pot,
Papa looking for a sweatshop to work in,
and the kids Sammy and Ethel saying Gee Whiz
and eating their first bananas that tasted funny to them.

They settle in a cold-water flat
(nine dollars a month in those days—
now ninety if you qualify professional)
in a tenement that is multiracial rather than interracial
(meaning No Colored),
with an Irish woman living on the top floor
who had two sons, one a gangster and the other a cop,
her sorrow being that she didn't have a third to be a priest
and that way cover all possible professions;
and a German couple running the candy store downstairs,
who later go back to Germany with photographs of the neighborhood
and a complete list of the Jews there for the master file.

Life goes on, the good times and the bad:
Papa right away organizes the workers
and they go out on strike.
Little Sammy reads Tolstoy and Marx
and becomes a hothead radical and argues.
He goes around singing:

The hat worker's union
is a fascist union,
The hat worker's union
is a fascist union,
they preach socialism
but they practice fascism
to make the world better for the boss class.

At last George our hero is born, the fruit of freedom,
and settles down to his music lessons,
with the family standing around beaming
as he plays 'Minuet in G.'

Before long he is sneaking out of the house nights.
For crime? No. After girls? No,
he was playing piano for the silent films.
(What am I saying 'films,' nowadays they're films,
then they were movies.)

His father is furious when he learns about this:
George was only supposed to play religious music.
Sammy and Ethel were the secular wing of the family.
What was all this about jazz? That is music?

But playing organ in the synagogue one Friday night
George fell to dreaming and broke into ragtime,
and pretty soon all those old bearded Jews
are tapping their feet and smiling.
So George is kicked out into the hard world
with no blessing from his father except, 'You bum you,'
and makes the rounds of Tin Pan Alley.

Sammy, the oldest, gives up Karl Marx
and marries a merchant's ugly daughter
and gets rich working in the business
and moves away to a better-class neighborhood like Brooklyn.

Little Ethel is growing up smart too,
and the struggling student who loves her
can hardly offer her the good times she had a right to
with her face and figure.
Pretty soon she is being driven home in a long limousine with curtains.
It pulled right up to the tenement door
with the neighbors staring from stoops and windows.
She drew a heavy veil over her face
and went up to the kitchen
where Ma turns a stone face to the stove.

'Look at you Ma,' she says, 'you're only forty.
I don't want that to happen to me,
so I'm getting out of this dump.'
And she goes off to be kept by the Irish woman's gangster son,
who was making a million as a bootlegger.

George writes song hit after song hit
and meets Alexis Smith, a noble society woman with ice-blond hair,
who comes to his mother's house Friday nights
for the chicken soup and matzoh balls.
'You know Mother Bernstein, uptown we don't get food like this.'
Anyway her French chef doesn't make it.
After about ten years' engagement
she gives George up because
she doesn't want to stand in the way of his music.

Meanwhile Ethel has been secretly putting her student through school
and paying the doctor bills for his crippled sister,
who wants to be a ballet dancer if the operation turns out successful.

But her gangster finally gets killed by his cop brother
when he comes, a fugitive, to his mother's kitchen door
and begs her to hide him from the law, and she says no.
He is then shot dead on the fire escape in the searchlights,
his blood dripping down on the elegant clothes
of Ethel weeping in the street below.

So Ethel comes back home and marries her student, now a lawyer,
but their first child is born dead
and the doctor says she will never be able to have another,
because of her past sins.

More success for George, in London and Paris,
command performance, cheering crowds, gray hair:
Tonight his great musical *Lower East Side* is opening
and it looks like a hit.
His leading lady loves him
but he always says, 'As long as momma is alive ...'

The curtain is about to go up on Act Three.
A telegram comes, his manager reads it:
PAPA HIT BY CAR, MOMMA HAS HEART ATTACK, COME AT ONCE. ETHEL.
The manager asks himself, Should I show it to him now?
George is standing in the wings ready to go out
and sing the finale himself.
His manager gives him the telegram.
He reads it as the orchestra plays the introduction, his cue.
The orchestra plays it again, and the audience starts murmuring.
George staggers out on stage humbly,
a small man with tears in his eyes,
holding the fatal telegram in his hand;
and looking up at the brightest light
that makes the tears click down his cheeks like diamonds,
he sings his big song.

Thunderous applause, cheers, rave reviews:
He takes the train to the funeral.
Standing around the grave, Ethel, Sammy, himself,
all grown up now, well dressed.
The old kitchen, oilcloth table top,
mama's pot she could always make soup in just by adding water.
What happened? Where did we go wrong?
Ethel and Sammy go away in their limousines:

'Got to be running along, kid. Call me sometime.'
Alone. Mama, Papa, where are you?
Tears. A hand falls on his shoulder. His manager.
He looks long into the loving eyes. He decides.

He puts a coin in the pay phone: 'Darling,
I want you to marry me, I love you.'
He is calling his leading lady, who has long loved him.
'Oh George,' she breathes, as the music
whirls up into a symphony of hearts full of happiness,
and the lovers rush into each other's arms in the middle of the stage
to sing the grand finale:
LOWER EAST SIDE.

VINCENT BUCKLEY

1925–1988

475 *Teaching German Literature*

I TEACH German literature, and this is how it goes:
Schiller, Böll and Hölderlin, and everybody knows
that Bertolt turned on Thomas Mann and punched him on the nose,
and Goethe married Clara Kronk and Clara married Wagner.

Hardy is a proximist and Philip Larkin hates
walking past the neighbour's children hanging on the gates.
Keats was orphaned, Donne was bent, and Shelley went out drowning,
and Wordsworth married Sara Dronk and Sara married Browning.

T. S. Eliot, marvellous boy, grew up with a mitre;
even in dark Russell Square you wouldn't find much brighter,
Fürster Rilke knew a duchess, Trudi von Bachbeiter,
and Mara married Pablo Yeats and Pablo never married.

The greats of German literature are in the dressing room,
Lotte Lenya, Suky Tawdry, fending off life's gloom,
Hans Otto Eller Manzenberger smiling at his poem,
and Clara married Leslie Liszt and Leslie married Magna.

And Goethe married Clara Kronk and Clara married Wagner.

KENNETH KOCH

1925–

476

from *Fresh Air*

SUPPOSING that one walks out into the air
On a fresh spring day and has the misfortune
To encounter an article on modern poetry
In *New World Writing*, or has the misfortune
To see some examples of some of the poetry
Written by the men with their eyes on the myth
And the Missus and the midterms, in the *Hudson Review*,
Or, if one is abroad, in *Botteghe Oscure*,
Or indeed in *Encounter*, what is one to do
With the rest of one's day that lies blasted to ruins
All bluely about one, what is one to do?
Oh surely one cannot complain to the President,
Nor even to the deans of Columbia College,
Nor to T. S. Eliot, nor to Ezra Pound,
And supposing one writes to the Princess Caetani,
'Your poets are awful!' what good would it do?
And supposing one goes to the *Hudson Review*
With a package of matches and sets fire to the building?
One ends up in prison with trial subscriptions
To the *Partisan*, *Sewanee*, and *Kenyon Review*!

A. R. AMMONS

1926–

477

Their Sex Life

ONE failure on
Top of another

478

Coming Right Up

ONE can't
have it

both ways
and both

ways is
the only

way I
want it.

NISSIM EZEKIEL

1926–

479

Goodbye Party for Miss Pushpa T. S.

FRIENDS,
our dear sister
is departing for foreign
in two three days,
and
we are meeting today
to wish her bon voyage.

You are all knowing, friends,
what sweetness is in Miss Pushpa.
I don't mean only external sweetness
but internal sweetness.
Miss Pushpa is smiling and smiling
even for no reason
but simply because she is feeling.

Miss Pushpa is coming
from very high family.
Her father was renowned advocate
in Bulsar or Surat,
I am not remembering now which place.

Surat? Ah, yes,
once only I stayed in Surat
with family members
of my uncle's very old friend,
his wife was cooking nicely . . .
that was long time ago.

Coming back to Miss Pushpa
she is most popular lady
with men also and ladies also.
Whenever I asked her to do anything,
she was saying, 'Just now only
I will do it.' That is showing
good spirit. I am always
appreciating the good spirit.
Pushpa Miss is never saying no.
Whatever I or anybody is asking
she is always saying yes,
and today she is going
to improve her prospect,
and we are wishing her bon voyage.

Now I ask other speakers to speak,
and afterwards Miss Pushpa
will do summing up.

480 from *Songs for Nandu Bhende*

(i)

Song to be Shouted Out

I COME home in the evening
and my wife shouts at me:
Did you post that letter?
Did you make that telephone call?
Did you pay that bill?
What do you do all day?

I come home in the evening
and my wife shouts at me:
Did you bank that cheque?
Did you buy those tickets?
Did you ask if cheese is in stock or not?
What do you do all day?

Shout at me, woman!
Pull me up for this and that.
You're right and I'm wrong.
This is not an excuse,
it's only a song.
It's good for my soul
to be shouted at.
Shout at me, woman!
What else are wives for?

(ii)

Family

ALL of us are sick, Sir,
not just the eldest daughter,
no, nor the younger one.
It's not the son, not the son,
It's all of us who need you, Sir
Psy—chi—a—trist!

Should we take to meditation,
transcendental, any other?
Should we take to Zen?
We cannot find our roots here,
don't know where to go, Sir,
don't know what to do, Sir,
need a Guru, need a God.
All of us are sick, Sir.

Time is ripe for Sai Baba.
Time is ripe for Muktananda.
Let father go to Rajneesh Ashram.
Let mother go to Gita classes.
What we need is meditation.
Need to find our roots, Sir.
All of us are sick, Sir.

481 from *Poems in the Greek Anthology Mode*

WHEN the female railway clerk
Received an offer of marriage
From her neighbour the customs clerk,
She told him to apply in triplicate,
And he did.

EDWARD GOREY

1926–

482

Limericks

(i)

FROM Number Nine, Penwiper Mews,
There is really abominable news:
They've discovered a head
In the box for the bread,
But nobody seems to know whose.

(ii)

SOME Harvard men, stalwart and hairy,
Drank up several bottles of sherry;
In the Yard around three
They were shrieking with glee:
'Come on out, we are burning a fairy!'

(iii)

TO his club-footed child said Lord Stipple,
As he poured his post-prandial tipple:
'Your mother's behaviour
Gave pain to Our Saviour
And that's why He made you a cripple.'

(iv)

FROM the bathing-machine came a din
As of jollification within;
It was heard far and wide
And the incoming tide
Had a definite flavour of gin.

TOM LEHRER

1928–

483

Wernher von Braun

GATHER round while I sing you of Wernher von Braun,
A man whose allegiance is ruled by expedience.
Call him a Nazi, he won't even frown,
'Nazi, Shmazi,' says Wernher von Braun.
  Don't say that he's hypocritical,
  Say rather that he's apolitical.
'Once the rockets are up, who cares where they come down?
That's not my department,' says Wernher von Braun.

Some have harsh words for this man of renown,
But some think our attitude should be one of gratitude,
Like the widows and cripples in old London town
Who owe their large pensions to Wernher von Braun.
  You too may be a big hero,
  Once you've learned to count backwards to zero.
'In German oder English I know how to count down,
Und I'm learning Chinese,' says Wernher von Braun.

DONALD HALL

1928–

484

Woolworth's

MY whole life has led me here.

Daisies made out of resin,
hairnets and submarines,
sandwiches, diaries, green
garden chairs,
and a thousand boxes of cough drops.

Three hundred years ago I was hedging
and ditching in Devon.
I lacked freedom of worship,
and freedom to trade molasses
for rum, for slaves, for molasses.

'I will sail to Massachusetts
to build the Kingdom
of Heaven on Earth!'

The side of a hill
swung open.
It was Woolworth's!

I followed this vision to Boston.

BRUCE BEAVER

1928–

485

Folk Song

O I'M off to Hullaboola where the climate's never cooler
Than a ringside seat in Hell. They're growing corn there
That pops the while it's growing, and the reason why I'm going
Is because I hate the name and wasn't born there.

So I'm leaving kin and kith for this never-never myth
Where Matilda warbles waltzes till she stutters.
Where the dinkum bunyips leer from the billabongs of beer
And the Clancys overflow into the gutters.

Clancys] 'Clancy of the Overflow', a poem by A. B. ('Banjo') Paterson

X. J. KENNEDY

1929–

486 from *Emily Dickinson in Southern California*

I CALLED one day—on Eden's strand
But did not find her—Home—
Surfboarders triumphed in—in Waves—
Archangels of the Foam—

I walked a pace—I tripped across
Browned couples—in cahoots—
No more than Tides need shells to fill
Did they need—bathing suits—

From low boughs—that the Sun kist—hung
A Fruit to taste—at will—
October rustled but—Mankind
Seemed elsewhere gone—to Fall—

487 *To Someone who Insisted I Look Up Someone*

I RANG them up while touring Timbuctoo,
Those bosom chums to whom you're known as '*Who?*'

RAY MATHEW

1929–

488 *Poem in Time of Winter*

MY head is unhappy,
My heart is like lead,
My chilblains are itchy,
I'd rather be dead.
My heart's like a horse-shoe,
And I never have luck,
But I don't give a damn,
I don't give a river,
I don't give a duck.

My girl's got a temper,
Her mother's a dog,
The pictures cost money,
She eats like a hog.
We ought to be kissing,
And I haven't the pluck,
But I don't give a damn,
I don't give a river,
I don't give a duck.

My friends are dying,
The happy are sad,
Are twisted with illness,
The good go mad.
I go to the church,
And they ask for the buck,
But I don't give a damn,
I don't give a river,
I don't give a duck.

The rain keeps raining,
The wet comes down,
It's so grey and horrible
To wait and drown
That I'd buy a car,
Or I'd thumb a truck—
But I don't give a damn,
I don't give a river,
I don't give a duck.

CONNIE BENSLEY

1929–

489 *Bloomsbury Snapshot*

VIRGINIA's writing her diary,
Vanessa is shelling the peas,
And Carrington's there, hiding under her hair,
And squinting, and painting the trees.

Well Maynard is smiling at Duncan,
A little to Lytton's distress,
But Ralph's lying down with a terrible frown
For he'd rather be back in the mess.

There's Ottoline, planning a party—
But Leonard's impassive as stone:
He knows that they'll all sit around in deck chairs,
Discussing their own and each others' affaires,
And forming, perhaps, into new sets of pairs:
And oh, how the bookshelves will groan.

490

One's Correspondence

I WROTE to you to say that I'd be there
but lost the letter giving your address
and now I cannot find it anywhere.

Although I've searched until I'm in despair,
what's worrying me most is, I confess,
I wrote to you to say that I'd be there.

It came first thing on Tuesday (to be fair
the breakfast table was in quite a mess)
and now I cannot find it anywhere.

I think you said you lived in Berkeley Square
or did you say you'd moved to Inverness?
I wrote to you to say that I'd be there.

Where parties are concerned, you have a flair.
The letter said: 'Please come in fancy dress,'
and now I cannot find it anywhere.

I'm sure I wrote a note but couldn't swear
to posting it: this is an SOS—
I wrote to you to say that I'd be there
and now I cannot find it anywhere.

PETER PORTER

1929–

491

Japanese Jokes

In his winged collar
he flew. The nation wanted
peace. Our Perseus!

William Blake, William
Blake, William Blake, William Blake,
say it and feel new!

Love without sex is
still the most efficient form
of hell known to man.

A professional
is one who believes he has
invented breathing.

The Creation had
to find room for the exper-
imental novel.

When daffodils be-
gin to peer: watch out, para-
noia's round the bend.

I get out of bed
and say goodbye to people
I won't meet again.

I sit and worry
about money who very
soon will have to die.

I consider it
my duty to be old hat
so you can hate me.

I am getting fat
and unattractive but so
much nicer to know.

Somewhere at the heart
of the universe sounds the
true mystic note: Me.

492 from *The Sanitized Sonnets*

NOW it's in all the novels, what's pornography to do?
Stay home where it's always been—in the mind.
It's always been easier to wank than to grind,
yet love is possible, palpable and happens to you.

It's nice to have someone say thank you afterwards
goes the old joke. But are the manual writers
right, are masturbators nail biters?
(Even the Freudians are anti, albeit in long words.)

Don't burn *Office Frolics* and *I've a Whip in my Valise*;
in other disciplines the paradis artificiel
is considered high art and not mental disease

and if your mind arranges tableaux with girls—
e.g. strip poker with big-breasted Annabel—
it's a sign the world's imperfect and needs miracles.

JOHN HOLLANDER

1929–

493 *Historical Reflections*

HIGGLEDY-PIGGLEDY,
Benjamin Harrison,
Twenty-third President,
Was, and, as such,

Served between Clevelands, and
Save for this trivial
Idiosyncracy,
Didn't do much.

494 *No Foundation*

HIGGLEDY-PIGGLEDY
John Simon Guggenheim,
Honored wherever the
Muses collect,

Save in the studies (like
Mine) which have suffered his
Unjustifiable,
Shocking neglect.

U. A. FANTHORPE

1929–

495 *You Will Be Hearing from us Shortly*

YOU feel adequate to the demands of this position?
What qualities do you feel you
Personally have to offer?

Ah

Let us consider your application form.
Your qualifications, though impressive, are
Not, we must admit, precisely what
We had in mind. Would you care
To defend their relevance?

Indeed

Now your age. Perhaps you feel able
To make your own comment about that,
Too? We are conscious ourselves
Of the need for a candidate with precisely
The right degree of immaturity.

So glad we agree

And now a delicate matter: your looks.
You do appreciate this work involves
Contact with the actual public? Might they,
Perhaps, find your appearance
Disturbing?

Quite so

And your accent. That is the way
You have always spoken, is it? What
Of your education? Were
You educated? We mean, of course,
Where were you educated?
And how
Much of a handicap is that to you,
Would you say?

Married, children,
We see. The usual dubious
Desire to perpetuate what had better
Not have happened at all. We do not
Ask what domestic disasters shimmer
Behind that vaguely unsuitable address.

And you were born—?

Yes. Pity.

So glad we agree.

ANTHONY THWAITE

1930–

496 *Great Foreign Writer Visits Age-Old Temple, Greeted by Venerable Abbess, 1955*

Based on a transcript, 'Interview at Zenkoji Temple', from *Faulkner in Nagano* (Tokyo, 1956)

GFW I AM most honoured
to be received here
this afternoon.

VA We are very glad
that you came
despite the heat.

GFW I hear there are
many National Treasures
in this temple.

VA This building is itself
a National Treasure,
as is the Buddha
deified inside it.

GFW May I ask
how old
the temple is?

VA This temple
is one thousand four hundred
years old. I am the one hundred
and nineteenth abbess.

GFW In what dynasty
was the temple founded?

VA In the era
of the Emperor Kinmei,
when the Buddha came
to Japan . . . May I ask
your purpose in coming
to Japan?

GFW I came to Japan
to know more about
the Japanese people
and Japanese culture,
of which we know something
in my country, and admire.

VA And may I ask how long
you intend to stay?

GFW Three weeks.

VA In regard to religion,
are you studying Christianity
or Buddhism primarily?

GFW I am interested in all religions
as a form of man's behaviour.

VA Is this your first
visit to Japan?

GFW Yes, but I have known
Japanese history
Japanese art
and Japanese literature
a long time.

VA I feel very much assured
that you have so much understanding
towards these things.

GFW Thank you very much.
I wish more people from my country
could know your people
and your country.

VA Can you eat
this kind of
Japanese cake?

GFW I am sure I can
because I like
all Japanese food.

VA Are you giving a lecture
or something?

GFW I am attending a seminar
on our country's literature.

VA It must be very trying
in this hot season.

GFW No, this season
is like the season at my home—
very pleasant.

VA Have you a message
for our youth? For the world?
What is your impression
of our women? Of God?
May I ask
what tobacco you smoke?

GFW To thine own self be true.
May peace prevail.
Very beautiful.
The same to all but
called by different names.
A blend I have made up.

VA Thank you very much for coming.

GFW Thank you for all your trouble.

Together Thank you. Thank you. Thank you.

GFW *Arigato . . .*
(Did I get that right?)

BRUCE DAWE

1930–

497 *Weapons Training*

AND when I say eyes right I want to hear
those eyeballs click and the gentle pitter-patter
of falling dandruff   you there what's the matter
why are you looking at me   are you a queer?
look to your front   if you had one more brain
it'd be lonely   what are you laughing at
you in the back row with the unsightly fat
between your elephant ears   open that drain
you call a mind and listen   remember first
the cockpit drill when you go down   be sure
the old crown-jewels are safely tucked away
        what could be more
distressing than to hold off with a burst
from your trusty weapon a mob of the little yellows
only to find back home because of your position
your chances of turning the key in the ignition
considerably reduced?   allright now suppose
for the sake of argument you've got
a number-one blockage and a brand-new pack
of Charlies are coming at you   you can smell their rotten
        fish-sauce breath hot on the back
of your stupid neck allright now what
are you going to do about it?   that's right grab and check
the magazine man it's not a woman's tit
worse luck or you'd be set   too late you nit
they're on you and your tripes are round your neck
you've copped the bloody lot just like I said
and you know what you are?   you're dead dead dead

R. J. P. HEWISON, GERRY HAMILL, and GERARD BENSON

498

Limericks

(i)

SAID an erudite sinologue: 'How
Shall I try to describe to you Tao?
 It is come, it is go,
 It is yes, it is no,
Yet it's neither—you understand now?'

R. J. P. HEWISON

(ii)

'IF you're aristocratic,' said Nietzsche,
'It's thumbs up, you're OK. Pleased to mietzsche.
 If you're working-class bores,
 It's thumbs down and up yours!
If you don't know your place, then I'll tietzsche.'

GERRY HAMILL

(iii)

THERE once was a bard of Hong Kong
Who thought limericks went on too long.

GERARD BENSON

P. J. KAVANAGH

1931–

499

Goldie Sapiens

WHEN Goldie the golden eagle escaped from the Zoo
All the world went to Regent's Park and we went too.
There he was, with an air of depression, a sooty hunch,
Digesting the grey-eyed merganser he had for lunch.
Under him children and coppers and mothers and fathers
And bare-kneed ornithologists with cameras
Hanging down to their ankles and lovers and others

Peeling damp cellophane from sandwiches stand and wait.
While running around in sad moustaches Keepers,
Hopelessly, like H. M. Bateman characters,
Shoo Pekes away from buckets of eagle bait.
Really, this bird was a National Occasion!
The Evening Standard published an hourly bulletin
As though it was getting in training for Sir Winston.
And none of us knew what we most wanted to see,
The Keepers allowed to go home or the bird to go free.
There was an appalling sense of a happy ending too—
Goldie was free but he kept an eye cocked on his Zoo.
Just then there started up where Goldie was,
A thrush fit to burst but we didn't listen because
We were enjoying the sight we'd come to see—
The only free eagle in captivity.
Later that evening the Nation breathed a sigh.
Goldie like us, Goldie the human and sage,
With tail between talons, had lolloped back to the cage.

E. WILLIAM SEAMAN and ERIC SALZMAN

500 *Double Dactyls*

(i)

HIGGLEDY-PIGGLEDY
Ludwig van Beethoven
Bored by requests for some
Music to hum,

Finally answered with
Oversimplicity,
'Here's my Fifth Symphony:
Duh, duh, duh, DUM!'

E. WILLIAM SEAMAN

(ii)

MOISHELE, Moishele
Moses Maimonides,
Bowing and chanting 'Bar-
Uch Adonai,'

Picked up the telephone
Anachronistically,
Ordered pastrami with
Mustard on rye.

ERIC SALZMAN

GEORGE STARBUCK

1931–

501

SAID
Agatha Christie to
E. Phillips Oppenheim,
'Who is this Hemingway,
Who is this Proust?

'Who is this Vladimir
Whatchamacallum, this
Neopostrealist
Rabble?' she groused.

JOHN UPDIKE

1932–

502

V. B. Nimble, V. B. Quick

Science, Pure and Applied, by V. B. Wigglesworth,[1] FRS, Quick Professor of Biology in the University of Cambridge.

—*a talk listed in the* BBC RADIO TIMES

V. B. WIGGLESWORTH wakes at noon,
Washes, shaves, and very soon
Is at the lab; he reads his mail,
Tweaks a tadpole by the tail,
Undoes his coat, removes his hat,
Dips a spider in a vat

Of alkaline, phones the press,
Tells them he is FRS,
Subdivides six protocells,
Kills a rat by ringing bells,
Writes a treatise, edits two
Symposia on 'Will Man Do?,'
Gives a lecture, audits three,
Has the Sperm Club in for tea,
Pensions off an aging spore,
Cracks a test tube, takes some pure
Science and applies it, finds
His hat, adjusts it, pulls the blinds,
Instructs the jellyfish to spawn,
And, by one o'clock, is gone.

[1] Author's Note: The V. B. Wigglesworth in this poem bears no resemblance whatever to the real V. B. Wigglesworth.

503 *The Newlyweds*

After a one-day honeymoon, the Fishers rushed off to a soft drink bottlers convention, then on to a ball game, a TV rehearsal and a movie preview.

—*Life*

'WE'RE married,' said Eddie.
Said Debbie, 'Incredi-

ble! When is our honey-
moon?' 'Over and done,' he

replied. 'Feeling logy?
Drink Coke.' 'Look at Yogi
go!' Debbie cried. 'Groovy!'
'Rehearsal?' 'The movie.'

'Some weddie,' said Debbie.
Said Eddie, 'Yeah, mebbe.'

504 *On the Inclusion of Miniature Dinosaurs in Breakfast Cereal Boxes*

A POST-HISTORIC herbivore,
I come to breakfast looking for
A bite. Behind the box of Brex
I find *Tyrannosaurus rex*.

And lo! beyond the Sugar Pops,
An acetate *Triceratops*.
And here! across the Shredded Wheat,
The spoor of *Brontosaurus* feet.

Too unawake to dwell upon
A model of *Iguanodon*,
I hide within the Raisin Bran;
And thus begins the dawn of *Man*.

505 *In extremis*

I SAW my toes the other day.
I hadn't looked at them for months.
Indeed, they might have passed away.
And yet they were my best friends once.

When I was small, I knew them well.
I counted on them up to ten
And put them in my mouth to tell
The larger from the lesser. Then

I loved them better than my ears,
My elbows, adenoids, and heart.
But with the swelling of the years
We drifted, toes and I, apart.

Now, gnarled and pale, each said, *j'accuse!*—
I hid them quickly in my shoes.

GEORGE MACBETH

1932–1992

506

The Orange Poem

NOT very long ago
One morning
I sat in my orange room
With my orange pencil
Eating an orange.

This,
I began to write,
Is the orange poem.
I shall become known
As the orange poet

For inventing
And first writing
The original
Perfect
And now famous

ORANGE POEM
Which this is.

Having written which
In my orange room
With my orange pencil

I turned over a new leaf
Which this is.
Meanwhile,
Inside the orange poem
A small man

With an orange pencil
Sat in an orange room
Eating an orange.
This, he began to write,
Is the orange poem.

ADRIAN MITCHELL

1932–

507

Icarus Schmicarus

If you never spend your money
you know you'll always have some cash.
If you stay cool and never burn
you'll never turn to ash.
If you lick the boots that kick you
then you'll never feel the lash,
and if you crawl along the ground
at least you'll never crash.
So why why why—
WHAT MADE YOU THINK YOU COULD FLY?

SHEL SILVERSTEIN

1932–

508

Slithergadee

THE Slithergadee has crawled out of the sea.
He may catch all the others, but he won't catch me.
No you won't catch me, old Slithergadee,
You may catch all the others, but you wo——

509

Jimmy Jet and his TV Set

I'LL tell you the story of Jimmy Jet—
And you know what I tell you is true.
He loved to watch his TV set
Almost as much as you.

He watched all day, he watched all night
Till he grew pale and lean,
From 'The Early Show' to 'The Late Late Show'
And all the shows between.

He watched till his eyes were frozen wide,
And his bottom grew into his chair.
And his chin turned into a tuning dial,
And antennae grew out of his hair.

And his brains turned into TV tubes,
And his face to a TV screen.
And two knobs saying 'VERT.' and 'HORIZ.'
Grew where his ears had been.

And he grew a plug that looked like a tail
So we plugged in little Jim.
And now instead of him watching TV
We all sit around and watch him.

JAMES SIMMONS

1933–

510 *The Pleasant Joys of Brotherhood*

(*To the tune of 'My Lagan Love'*)

I LOVE the small hours of the night
when I sit up alone.
I love my family, wife and friends.
I love them and they're gone.
A glass of Power's, a well-slacked fire,
I wind the gramophone.
The pleasant joys of brotherhood
I savour on my own.

An instrument to play upon,
books, records on the shelf,
and albums crammed with photographs:
I *céilí* by myself.
I drink to passion, drink to peace,
the silent telephone.
The pleasant joys of brotherhood
I savour on my own.

Céilí] friendly visit, social evening

BARRY HUMPHRIES

1934–

511

Edna's Alphabet

A IS for apron in plastic or cloth,
It protects all your dresses from hot fat or broth.

B is for bathroom where dirt I deplore,
You ought to be able to eat off the floor.

C stands for cuppa, refreshing and hot,
It won't taste the same if you don't warm the pot.

D is for dripping which drips from the roast,
Essential for pastry, delicious on toast.

E stands for eggs which the kiddies adore.
Remember preserving them during the War?

F is for fridge ideal when it's summery
For keeping fresh fruit, fish, flan, fudge and flummery.

G is for gully trap where old tea leaves cling;
When rinsing it out hang on to your ring.

H stands for hygiene and so I say please,
Don't make of your kitchen a bed of disease.

I is for invalid, give them a treat.
Remember the sick tray and a hottie for their feet.

J stands for jug which we plug in a point,
It's also for jam, jelly, junket and joint.

K is for kitchen the hub of the home
I'm sitting there now writing this poem.

L is for lino, Laxettes and lav,
Three little essentials a family must have.

M is for Mother from whom we all learn.
I was learning right up till the time of her turn.

N is for Norm, that husband of mine,
He's giving my passage a wonderful shine.

O is for oven with its warm friendly glow—
Remind me to turn my bottom to low.

P is for pride in this great land of ours,
Our culture, our history, I could go on for hours.

Q stands for quantity, give nice generous serves;
Folk who are stingy get on my nerves.

R is for recipes I give mine with pride,
Those in my home have been thoroughly tried.

S is for sponge cake, of good cooks a sign,
Though I say it myself there are none to beat mine.

T stands for toddler, the joy of our lives,
Though it breaks your heart when they're covered with hives.

U for utensil spotlessly clean,
Scour straight after use for perfect hygiene.

V stands for vegies; eat some each day—
The vitamins there keep the doctor away.

W is for wash-day though Valmai may scoff,
For undies (personals) I still use my copper and trough.

X is for Xmas the time of good cheer
Though I don't get a break with the family here.

Y stands for You Yangs, those mountains so far.
When we drove there young Kenny was sick in the car.

Z is for zest, the secret of life;
You'll skim through the days a good mother and wife.

CHRIS WALLACE-CRABBE

1934–

512 *Binary*

WHY does a cauliflower so much resemble a brain? All those pale curved protruberances and hillocks tease the mind into activity . . . at this point I randomly remember the complicated architecture of a particular gothic dream. But brains and cauliflowers, ah yes, is this the same kind of parallelism as that which holds between pine cones and pagodas? There we go again, seeking order or duplicity in the stubborn universe. We ask ourselves whether the resemblance between a rose and a cabbage is like that between a clipped hedge and a high tin loaf, coming up with no answer at all. The brain and the cauliflower continue to rise up on their cortices, bubbly fruits that they are. Moon goes around earth goes around sun, et cetera, analogues active everywhere. Echo redeems Narcissus, shadow is touching reflection. We ask ourselves what it all signifies. Somewhere, in shadow, aged sages debate such questions on a lawn all day, over their wine and bananas.

513 *Double Dactyl*

HIGGLEDY-PIGGLEDY
Thomas Stearns Eliot
Made a whole poem to
Carry one word.

What was it now? *Poly-*
Philoprogenitive.
I do not like it. I
Think it absurd.

FLEUR ADCOCK

1934–

514 *The Three-Toed Sloth*

THE three-toed sloth is the slowest creature we know
for its size. It spends its life hanging upside-down
from a branch, its baby nestling on its breast.
It never cleans itself, but lets fungus grow
on its fur. The grin it wears, like an idiot clown,
proclaims the joys of a life which is one long rest.

The three-toed sloth is content. It doesn't care.
It moves imperceptibly, like the laziest snail
you ever saw blown up to the size of a sheep.
Disguised as a grey-green bough it dangles there
in the steamy Amazon jungle. That long-drawn wail
is its slow-motion sneeze. Then it falls asleep.

One cannot but envy such torpor. Its top speed,
when rushing to save its young, is a dramatic
fourteen feet per minute, in a race with fate.
The puzzle is this, though: how did nature breed
a race so determinedly unenergetic?
What passion ever inspired a sloth to mate?

EDWARD BAUGH

1936–

515 *The Carpenter's Complaint*

NOW you think that is right, sah? Talk the truth.
The man was mi friend. *I* build it, *I*
Build the house that him live in; but now
That him dead, that marga-foot bwoy, him son,
Come say, him want a nice job for the coffin,

marga-foot] skinny

So him give it to *Mister* Belnavis to make—
That big-belly crook who don't know him arse
From a chisel, but because him is big-shot, because
Him make big-shot coffin, fi-him coffin must better
Than mine! Bwoy, it hot me, it hot me
For true. Fix we a nex' one, Miss Fergie—
That man coulda knock back him waters, you know sah!
I remember the day in this said-same bar
When him drink Old Brown and Coxs'n into
The ground, then stand up straight as a plumb-line
And keel him felt hat on him head and walk
Home cool, cool, cool. Dem was water-bird, brother!
Funeral? *Me*, sah? That bwoy have to learn
That a man have him pride. But bless mi days!
Good enough to make the house that him live in,
But not good enough to make him coffin!
I woulda do it for nutt'n, for nutt'n! The man
Was mi friend. Damn marga-foot bwoy.
Is university turn him fool. I tell you,
It burn me, it burn me for true!

JOHN FULLER (1937–) and JAMES FENTON (1949–)

516 *The Red Light District Nurse*

YOU'LL see me park my car upon
A double yellow line.
The wardens doff their hats at me
And there's no clamp or fine,
For if you put the clamp on me
You drive off in a hearse.
  Don't mess around,
  Don't press around
The red light district nurse.

fi-him] his *waters*] rum

They trained me on the Reeperbahn.
They passed me in Amsterdam.
They crammed me in a uniform
(And there's quite a lot to cram).
They crammed and crammed until they found
I'd scored a double First.
   Anatomy?
   What's that to me?
I'm the red light district nurse.

The Local Health Authority
Supplies me with high heels
For going down on pensioners
Between the meals on wheels,
For my money's where my mouth is,
I don't put it in my purse.
   I get the feel of it
   And make a meal of it,
The red light district nurse.

What's wrong with you? I'll tell you, son.
You drank a bit too much
And went to the tattooist's shop
And redesigned your crotch.
You wonder what your wife will think.
If she's like me, she'll burst.
   Don't pester me.
   Say Yes to me!
The red light district nurse.

Two lesbots, one banana:
You have to tug and tug.
It's terrible when the fuse explodes
In an ancient one-pin plug,
But I'm there on time if they call me
And I put them in reverse.
   I've been through it myself.
   (I still do it myself!)
I'm the red light district nurse.

There were three drunken Irishmen
Who called on me one day.
They weren't on the National Health and so
I had to make them pay.

It was a straight two tricks with the first two Micks.
With the Fenian it was Erse.
  It's a hat-trick
  For St Patrick
And the red light district nurse.

I've customers who come to me
For a thorough rectal search
And I've patients who are satisfied
With sauna, tea and birch,
But some can take it out of you
And a few I call perverse
  (Colostomy?
  Get lost Tommy!)
I'm the red light district nurse!

517 *Born Too Soon*

ALL the opportunities we miss by being born too soon!
All the pleasures of the evening lost by going to bed at noon!
Lydgate never peeled an orange, Langland never bit a Mars.
Chaucer might have felt quite raffish if he'd learned to smoke cigars.
  Born too soon, born too soon.

Bullough's *Sources* might have meant that Shakespeare's plots were more
    than cursory.
Marlowe never had the chance to bank a large Arts Council Bursary.
Herrick's sweet disorder counts for nothing in an age of zips.
Penicillin might have strengthened Davenant's relationships.
  Born too soon, born too soon.

Donne avoided death duties and Rochester a parking ticket.
Marvell would have simply *loved* a day or two of country cricket.
Richard Crashaw yearned in vain for something like a heart transplant.
Joseph Addison just missed becoming the first agony aunt.
  Born too soon, born too soon.

Poor old Milton, locked in darkness: not by bribery or stealth
Could he procure, like A. N. Wilson, glasses on the National Health,
Nor Pope, when furnishing his grotto, quantities of garden gnomes.
Swift's work suffers from his ignorance of missing chromosomes.
  Born too soon, born too soon.

Fanny Burney needed pretty coloured jackets from Virago.
Blake and Smart would have been quite at home in *Poetry* (Chicago).
Gray, if sex had been invented, would have found the knowledge bliss.
William Cowper needed several years of deep analysis.
Born too soon, born too soon.

Dorothy at Grasmere might have triumphed with a Hoovermatic,
William saved his quill enlarging snaps of Keswick in the attic.
Samuel needed fees for talking? Why not join the BBC?
Sara might have welcomed Sam, getting kicks from being three.
Born too soon, born too soon.

Shelley would have benefited from aerobics and from jogging,
Listerine have justified the urgency of Keats's snogging.
Byron never used a Pentax, posing Arab boys in Cairo
(More of Donny Johnny might have surfaced if he'd used a biro).
Born too soon, born too soon.

Branwell Brontë never had to face his sister's querying gaze
When another undergardener left the vicarage in a haze.
(Housebound Cathy found that Heathcliff somewhat cramped her social range:
'Wuthering Heights Two-Seven: Operator, give me Thrushcross Grange!')
Born too soon, born too soon.

Arnold missed the tape recorder, Tennyson the silver screen,
Robert Browning never saw a peepshow of the kind we mean.
Television would have ushered Dickens into *every* home.
Edward Lear's existence might have been transformed by shaving foam.
Born too soon, born too soon.

T. S. Eliot had (and missed) the chance to go and toss the caber
But he never saw his musical provide a float for Faber.
Hopkins could have read his stuff (with scansion marks) on Radio 3.
Rudyard Kipling could have eaten Mr Kipling cakes for tea.
Born too soon, born too soon.

Think of what we're losing now by having come to birth too soon:
All the future we imagine, haunting as a honeymoon.
You and us, like them, would gladly have performed it with impunity,
That delightful, stimulating, tragic, missing opportunity.
Born too soon, born too soon.

ROGER McGOUGH

1937–

518 from *Summer with Monika*

away from you
i feel a great emptiness
a gnawing loneliness

with you
i get that reassuring feeling
of wanting to escape

519 *40– Love*

MIDDLE	aged
couple	playing
ten	nis
when	the
game	ends
and	they
go	home
the	net
will	still
be	be
tween	them

520 *italic*

ONCE I LIVED IN CAPITALS
MY LIFE INTENSELY PHALLIC

but now i'm sadly lowercase
with the occasional *italic*

521

Survivor

EVERYDAY
I think about dying.
About disease, starvation,
violence, terrorism, war,
the end of the world.

It helps
keep my mind off things.

522

Poem with a Limp

WOKE up this morning with a
limp.
Was it from playing
football
In my dreams? Arthrite's first
arrow?
Polio? Muscular dystrophy? (A bit of
each?)

I staggered around the kitchen spilling
coffee
Before hobbling to the bank for
lire
For the holiday I knew I would not be
taking.
(For Portofino read Stoke
Mandeville.)

Confined to a wheelchair for the
remainder
Of my short and tragic life.
Wheeled
On stage to read my terse, honest
poems
Without a trace of bitterness. 'How
brave,
And smiling still, despite the
pain.'
Resigned now to a life of quiet
fortitude

I plan the nurses' audition.
Mid-afternoon
Sees me in the garden, sunning my
limp.

* * *

It feels a little easier now.
Perhaps a miracle is on its way?
(Lourdes, W11.)

By opening-time the cure is complete.
I rise from my deck-chair:
'Look, everybody, I can walk, I can walk.'

LES A. MURRAY

1938–

523 *Hearing Impairment*

HEARING loss? Yes, loss is what we hear
who are starting to go deaf. Loss
trails a lot of weird puns in its wake, viz.
Dad's a real prism of the Left—
you'd like me to repeat that?
THE SAD SURREALISM OF THE DEAF.

It's mind over mutter at work
guessing half what the munglers are saying
and society's worse. Punchlines elude to you
as Henry Lawson and other touchy drinkers
have claimed. Asides, too, go pasture.
It's particularly nasty with a wether.

First you crane at people, face them
while you can still face them. But grudgually
you give up dinnier parties; you begin
to think about Beethoven; you Hanover

next visit here on silly Narda Fearing—I SAY
YOU CAN HAVE AN EXQUISITE EAR
AND STILL BE HARD OF HEARING.

It seems to be mainly speech, at first,
that escapes you—and that can be a rest,
the poor man's escape itch from Babel.
You can still hear a duck way upriver,
a lorry miles off on the highway. You
can still say boo to a goose and
read its curt yellow-lipped reply.
You can shout SING UP to a magpie,

but one day soon you must feel
the silent stopwatch chill your ear
in the doctor's rooms, and be wired
back into a slightly thinned world
with a faint plastic undertone to it
and, if the rumours are true, snatches
of static, music, police transmissions:
it's a BARF minor Car Fourteen prospect.

But maybe hearing aids are now perfect
and maybe it's not all that soon.
Sweet nothings in your ear are still sweet;
you've heard the human range by your age
and can follow most talk from memory;
the peace of the graveyard's well up
on that of the grave. And the world would
enjoy peace and birdsong for more moments

if you were head of government, enquiring
of an aide Why, Simpkins, do you tell me
a warrior is a ready flirt?
I might argue—and flowers keep blooming
as he swallows his larynx to shriek
our common mind-overloading sentence:
I'M SORRY, SIR, IT'S A RED ALERT!

CLIVE JAMES

1939–

524 *The Book of my Enemy Has Been Remaindered*

THE book of my enemy has been remaindered
And I am pleased.
In vast quantities it has been remaindered.
Like a van-load of counterfeit that has been seized
And sits in piles in a police warehouse,
My enemy's much-prized effort sits in piles
In the kind of bookshop where remaindering occurs.
Great, square stacks of rejected books and, between them, aisles
One passes down reflecting on life's vanities,
Pausing to remember all those thoughtful reviews
Lavished to no avail upon one's enemy's book—
For behold, here is that book
Among these ranks and banks of duds,
These ponderous and seemingly irreducible cairns
Of complete stiffs.

The book of my enemy has been remaindered
And I rejoice.
It has gone with bowed head like a defeated legion
Beneath the yoke.
What avail him now his awards and prizes,
The praise expended upon his meticulous technique,
His individual new voice?
Knocked into the middle of next week
His brainchild now consorts with the bad buys,
The sinkers, clinkers, dogs and dregs,
The Edsels of the world of movable type,
The bummers that no amount of hype could shift,
The unbudgeable turkeys.

Yea, his slim volume with its understated wrapper
Bathes in the glare of the brightly jacketed *Hitler's War Machine*,
His unmistakably individual new voice
Shares the same scrapyard with a forlorn skyscraper
Of *The Kung-Fu Cookbook*,

His honesty, proclaimed by himself and believed in by others,
His renowned abhorrence of all posturing and pretence,
Is there with *Pertwee's Promenades and Pierrots—*
One Hundred Years of Seaside Entertainment,
And (oh, this above all) his sensibility,
His sensibility and its hair-like filaments,
His delicate, quivering sensibility is now as one
With *Barbara Windsor's Book of Boobs*,
A volume graced by the descriptive rubric
'My boobs will give everyone hours of fun.'

Soon now a book of mine could be remaindered also,
Though not to the monumental extent
In which the chastisement of remaindering has been meted out
To the book of my enemy,
Since in the case of my own book it will be due
To a miscalculated print run, a marketing error—
Nothing to do with merit.
But just supposing that such an event should hold
Some slight element of sadness, it will be offset
By the memory of this sweet moment.
Chill the champagne and polish the crystal goblets!
The book of my enemy has been remaindered
And I am glad.

525

from *Robert Lowell's Notebook*

(i)

Notes for a Sonnet

STALLED before my metal shaving mirror
With a locked razor in my hand I think of Tantalus
Whose lake retreats below the fractured lower lip
Of my will. Splinter the groined eyeballs of our sin,
Ford Madox Ford: you on the Quaker golf-course
In Nantucket double-dealt your practised lies
Flattering the others and me we'd be great poets.
How wrong you were in their case. And now Nixon,
Nixon rolls in the harpoon ropes and smashes with his flukes
The frail gunwales of our beleaguered art. What
Else remains now but your England, Ford? There's not
Much Lowell-praise left in Mailer but could be Alvarez

Might still write that book. In the skunk-hour
My mind's not right. But there will be
Fifty-six new sonnets by tomorrow night.

(ii)

Revised Notes for a Sonnet

ON the steps of the Pentagon I tucked my skull
Well down between my knees, thinking of Cordell Hull
Cabot Lodge Van du Plessis Stuyvesant, our gardener,
Who'd stop me playing speedway in the red-and-rust
Model A Ford that got clapped out on Cape Cod
And wound up as a seed-shed. Oh my God, my God,
How this administration bleeds but will not die,
Hacking at the rib-cage of our art. You were wrong, R. P.
Blackmur. Some of the others had our insight, too:
Though I suppose I had endurance, toughness, faith,
Sensitivity, intelligence and talent. My mind's not right.
With groined, sinning eyeballs I write sonnets until dawn
Is published over London like a row of books by Faber—
Then shave myself with Uncle's full-dress sabre.

(iii)

Notes for a Revised Sonnet

SLICING my head off shaving I think of Charles I
Bowing to the groined eyeball of Cromwell's sinning will.
Think too of Orpheus, whose disembodied head
Dumped by the Bacchants floated singing in the river,
His love for Eurydice surviving her dumb move
By many sonnets. Decapitation wouldn't slow me down
By more than a hundred lines a day. R. P. and F. M. F.
Play eighteen holes together in my troubled mind,
Ford faking his card, Blackmur explicating his,
And what is love? John Berryman, if you'd had what it took
We could have both blown England open. Now, alone,
With a plush new set-up to move into and shake down,
I snow-job Stephen Spender while the liquor flows like lava
In the parlour of the Marchioness of Dufferin and Ava.

TOM DISCH

1940–

526 *Abecedary*

A IS an Apple, as everyone knows.
But B is a . . . What do you suppose?
A Bible? A Barber? A Banquet? A Bank?
No, B is this Boat, the night that it sank.
C is its Captain, and D is its Dory,
While E—But first let me tell you a story.
There once was an Eagle exceedingly proud
Who thought it would fly, in the Form of a cloud—
Yes, E is for Eagle, and F is for Form,
And G is the Grass that got wet in a storm
When the cloud that the Eagle unwisely became
Sprinkled our hero and all of his fame
Over ten acres of upland plateau.
So much for that story. Now H. Do you know?
H is the Hay that was made from the Grass,
And I's the Idea of going to Mass,
Which is something that only a Catholic would do.
Jews go to Synagogue. J is a Jew.
K is for Kitchen as well as for Kiss,
While L is for all of the black Licorice
You can eat in an hour without feeling ill.
M is for Millipede, Millet, and Mill.
The first is an insect, the second a grain,
The third grinds the second: it's hard to explain
Such a process to children who never have seen it—
So let's go to the country right now! Yes, I mean it.
We're leaving already, and N is the Night
We race through to reach it, while P is the Plight
Of the people (Remember?) who sailed in that Boat
That is still, by a miracle, somehow afloat!
(Oh dear, I've just noticed I've overlooked O:
O's an Omission and really should go
In that hole—do you see it?—between N and P.
No? It's not there now? Dear O, pardon me.)
Q is the Question of how far away
A person can travel in one single day,

And whether it's worth it, or might it be better
To just stay at home and write someone a letter?
R's are Relations, a regular swarm.
Now get out of the car—we've arrived at their farm!
S is the Sight of a Thanksgiving feast,
And T is the Turkey, which must weigh at least
Thirty pounds. U is Utopia. V . . .
V simply Vanishes—where, we can't see—
While W Waves from its Westernmost isle
And X lies exhausted, attempting to smile.
There are no letters left now but Y and then Z.
Y is for You, dear, and Z is for me.

527

Zewhyexary

Z IS the Zenith from which we decline,
While Y is your Yelp as you're twisting your spine.
X is for Xmas; the alternative
Is an X-ray that gives you just one year to live.
So three cheers for Santa, and onward to W.
W's Worry, but don't let it trouble you:
W easily might have been Worse.
V, unavoidably, has to be Verse.
U is Uncertainty. T is a Trial
At which every objection is met with denial.
S is a Sentence of 'Guilty as Charged.'
R is a Russian whose nose is enlarged
By inveterate drinking, while Q is the Quiet
That falls on a neighborhood after a riot.
P is a Pauper with nary a hope
Of lining his pockets or learning to cope.
O is an Organ transplanted in vain,
While N is the Number of 'Enemies Slain':
Three thousand three hundred and seventy-three.
If no one else wants it, could M be for Me?
No, M is reserved for a mad Millionaire,
And L is his Likewise, and goes to his heir.
K is a Kick in the seat of your pants,
And J is the Jury whose gross ignorance
Guaranteed the debacle referred to above.
I's the Inevitability of
Continued inflation and runaway crime,
So draw out your savings and have a good time.

H is your Heart at the moment it breaks,
And G is the Guile it initially takes
To pretend to believe that it someday will heal.
F is the strange Fascination we feel
For whatever's Evil—Yes, Evil is E—
And D is our Dread at the sight of a C,
Which is Corpse, as you've surely foreseen. B is Bone.
A could be anything. A is unknown.

BASIL RANSOME-DAVIES

1940–

528 *Raymond Chandler: The Big Sleep*

My name's Philip Marlowe, the chivalrous shamus.
I'm not very rich and I'm not very famous.
For twenty-five dollars a day and expenses,
I move in a sub-world of thieves, pimps and fences.
Now General Sternwood has just had a feeler
From Arthur Gwynn Geiger, the fag porno dealer.
It seems that young Carmen, the General's daughter,
Has run up some debts and this Geiger had bought a
Fat sheaf of her IOUs, hoping to blackmail
Old Sternwood, for whom I'm a hot-on-the-track male.
Still other complexities thicken the mystery,
All somehow connected with family history:
Like where's Rusty Regan? (The vanishing Paddy
Who's missed by the Sternwoods, and not only Daddy,
But also by Vivian, Carmen's big sister,
Since he was her husband—a hell of a twister!)
So the plot is bizarre, hence my *précis* is minimal,
But it focuses on a degenerate criminal:
This guy Eddie Mars, and his web of ill-doing,
That accounts for a lot of the conflicts ensuing.
And the corpses pile up, but the principal five are
The smut-pedlar, Geiger; a horny young driver,
Who worked for the Sternwoods; Joe Brody, a vicious
Small-timer, who got his for being ambitious;

A P.I. called Harry Jones, sent into limbo
(The poor sap had fallen for Brody's ex-bimbo)
By Mars' man Canino—remorseless, inhuman—
Whom I shot myself to ensure the denouement.
Plus it turns out that Regan's as dead as a gopher,
But I'm damned if I figured who murdered the chauffeur.

E. J. THRIBB

529 *Lines on the Award 'Pipe Man of the Year' to Magnus Magnusson*

So. Magnus
Magnusson.

You are the
'Pipeman of
The Year'.

Pipeman.
A curious title.

How do you
Win?

Surely not
Just by
Smoking a
Pipe?

There would
Have to be more
To it than that.
Or would there?

E. Jarvis Thribb (17)

E. J. Thribb] perennially youthful contributor to the satirical magazine *Private Eye*

530

In Memoriam Larry Parnes *('Mr Parnes Shillings and Pence')*

So. Farewell
Then
Larry Parnes

Pop impresario.

Yes, you
Discovered Dicky Pride,
Johnny Gentle, Vince
Eager, Marty Wilde
And Tommy Steele.

'I'll make you
A star.'

That was
Your catchphrase.

'Where's my
Money?'

That was
Theirs.

E. J. Thribb (17)

531

In Memoriam Krishna Menon

So. Farewell Krishna
Menon.

Krishna. That
Is familiar on
Account of the members of the
Exotic Religious sect with

Shaven heads and
Clashing Cymbals

Who can be seen
Dancing through the
Streets. A strange
Spectacle.

Whence do they come?
Whither do they go?

(Personally I would not
Like to have my
Head shaven when
The weather's like this.)

E. J. Thribb (17)

532 *In Memoriam Salvador Dali*

So.
Hello then
Dali.

You are
A
Fish.

Keith's Mum
Is
Melting.

E. J. Watch (17)

GEOFF PAGE

1940–

533 *In Dante's Hell*

In Dante's Hell
the talk goes on
of southern slopes
and vignerons.

They tread a separate
circle there,
the wine-snobs parched
in their despair.

The floors are wide
with stately tables
bearing wine
of peerless label;

the walls go up
in serried racks
of noble imports
free from tax;

the air is rich
with stilton, brie
and camembert
from Normandy.

Like inmates of
some dismal camp
they circle lit
by discreet lamps

and talk of all
the classic years
and cannot quite
hold back their tears

when reminiscing
on all those
varietal wines
with 'lovely nose',

the late-picked rieslings
'autumn-scented',
the heady moselles
cold-fermented,

the 'bigness' of
a certain red
and what might follow
it in bed.

Their terms for him
who set them here
are those they once
reserved for beer.

And as their looks
grow more accusing
they do not find
the wines 'amusing'

for Satan lets them
talk and think,
do everything
in fact but drink

the vintages
from his high store.
'To drink,' he says,
'would surely bore.

A lifetime spent
in love with talk—
what need is there
to draw the cork?'

And so they ferment
two by two
and pay Beelze-
bub his due.

Their talk is muted,
sad and dry
and not a little
anguished by

the upper circle
howls of pain
from puritans on
French champagne.

JOHN MOLE

1941–

534

from *Penny Toys*

(i)

The Musical Monkey

THE musical monkey is dressed like a flunkey
In bell-bottom trousers and little peaked cap;
His master the grinder could hardly be kinder
And everyone calls him an elegant chap.
But see how his face is a world of grimaces
Which might make us wonder and should give us pause;
Oh quaint little creature, oh great Mother Nature,
A dance for our penny, a fig for your laws.

(ii)

Song of the Hat-Raising Doll

I RAISE my hat
And lower it.
As I unwind
I slow a bit.
This life—
I make a go of it
But tick-tock time
I know of it.

Yes, tick-tock time
I know of it.
I fear the final
O of it,
But making
A brave show of it
I raise my hat
And lower it.

KIT WRIGHT

1944–

535 *How the Wild South East Was Lost*

SEE, I was raised on the wild side, border country,
Kent 'n' Surrey, a spit from the country line,
An' they bring me up in a prep school over the canyon:
Weren't no irregular verb I couldn't call mine.

Them days, I seen oldtimers set in the ranch-house
(Talkin' 'bout J. 'Boy' Hobbs and Pat C. Hendren)
Blow a man clean away with a Greek optative,
Scripture test, or a sprig o' that rho-do-dendron.

Hard pedallin' country, stranger, flint 'n' chalkface,
Evergreen needles, acorns an' beechmast shells,
But atop that old lone pine you could squint clean over
To the dome o' the Chamber o' Commerce in Tunbridge Wells.

Yep, I was raised in them changeable weather conditions:
I seen 'em, afternoon of a sunny dawn,
Clack up the deck chairs, bolt for the back French windows
When they bin drinkin' that strong tea on the lawn.

In a cloud o' pipesmoke rollin' there over the canyon,
Book-larned me up that Minor Scholarship stuff:
Bent my back to that in-between innings light roller
And life weren't easy. And that's why I'm so tough.

536 *Victorian Family Photograph*

HERE is the mother all boobed and bodicey
Who started the children upon their odyssey.

There sits the father stern as a rock
Who rules the world with his iron cock.

Those the two children white as mice
Who saw the ghost in the attic, twice.

And who are we to suppose this vignette
Not threaded with love like a string quartet?

537 *Unlikely Obbligato of Andersonstown*

'O SHORT shrift's the best shrift to give to this *Festschrift*!'
The poor old Professor moaned, citing reviews
Of the tome in his 'honour' but lighting upon her
Has changed his estrangement and made him enthuse:

'As right as a trivet and nearer than privet,
As white and as light as the snow milling down,
O she is the rarest and likewise the fairest
That ever went walking through Andersonstown.'

Dispraise universal at read-through, rehearsal
And scunnered short run meant the elderly ham
Was half suicidal but viewing his idol
Has swivelled his snivel to *Cherchez la Femme*.

'O hear my hosanna! Eyes blue as Fermanagh,
Where any old-stager might happily drown,
And bright as the sun on the water, that daughter,
The loveliest ever in Andersonstown.'

The stricken logician reviewed his position.
He couldn't decide if he was or was not,
With nothing to go on but so forth and so on
And so forth. Her beauty has taught him what's what.

'O through the mind's mazes, let ring out her praises,
Though bombs they go up as the bullets come down,
Her loveliness one shot that's cleaner than gunshot
That ever went flying through Andersonstown.'

So let it be re-capped. O never be knee-capped:
If you're not the queen of it, I am the clown.
Take care of your beauty: that's only your duty
To three poor sods stomping through Andersonstown.

538 *Underneath the Archers, or What's All This about Walter's Willy?*

EVERYONE's on about Walter's willy
  Down at the Bull tonight.
He's done Dan's sheep and he's done them silly—
He's had young Phil and his daughter's filly—
  And folk don't think it's right.

  Folk *know* it can't be right.

No, the chat's not prim and the chat's not proper,
  Down at the Bull tonight,
'Cos everyone's on about Walter's whopper
And telling tales of his terrible chopper—
  And folk don't think it's right.

Sid Perks has drained the bitter cup
  Down at the Bull tonight.
Can't stand . . . or sit . . . or speak . . . or sup . . .
Walter got him while bottling up—
  And folk don't think it's right.

  Folk *know* it can't be right.

He got poor Polly while drawing a cork
  Down at the Bull tonight.
And Doris is still too ill to talk—
And Mrs Perkins can hardly *walk*—
  And folk don't think it's right.

There's in-depth discussion of every facet,
  Down at the Bull tonight,
Of Walter's gigantic natural asset—
Carries as far as Penny Hasset—
  (Folk know *that* can't be right)

  Folk *know* it can't be right.

Poor old Dan's a broken man
  Down at the Bull tonight.
Got locked in the back of Walter's van
With its ghastly height, unearthly span—
  And folk don't think it's right.

The Archers] long-running rural soap opera on BBC radio

Found him alone in the woods on Sunday
   (Down at the Bull tonight),
Had him all day and most of Monday—
That was the end of poor Joe Grundy—
   Folk don't think it's right.

   Folk *know* it can't be right.

It wasn't a Gainsborough nor an El Greco
   (Down at the Bull tonight)
Brought dozens of coach-loads out for a dekko—
But a photo-fit in the *Borchester Echo*—
   Folk don't think it's right.

Nobody understands it fully,
   Down at the Bull tonight,
The monstrous *range* of it. Was it by pulley
It scaled Grey Gables and whopped Jack Woolley?
   Folk don't think it's right.

   Folk *know* it can't be right.

There's coaches come from Ware and Wigan,
   Down at the Bull tonight,
From Wales and Wallasey, out for a swig an'
A sizing-up of Walter's big 'un—
   And folk don't think it's right.

Yes, everyone's on about Walter's thuggery,
   Down at the Bull tonight,
His *cattle*-courting, his *sheep* skullduggery,
Piggery jiggery-pokery buggery—
   Folk don't think it's right.

Even the Vicar's been muttering, 'F*** it',
   Down at the Bull tonight,
'There's nowhere left he hasn't stuck it—
I *wish* old Walter would kick the bucket—
   He knows it can't be right!'

   Folk know it CAN'T BE RIGHT!

PAUL DURCAN

1944–

539 *Sister Agnes Writes to her Beloved Mother*

DEAR MOTHER, Thank you for the egg cosy;
Sister Alberta (from near Clonakilty)
Said it was the nicest, positively the nicest,
Egg cosy she had ever seen. Here
The big news is that Revd Mother is pregnant;
The whole convent is simply delighted;
We don't know who the lucky father is
But we have a shrewd idea who it might be:
Do you remember that Retreat Director
I wrote to you about?—The lovely old Jesuit
With a rosy nose—We think it was he—
So shy and retiring, just the type;
Fr P. J. Pegasus SJ.
Of course, it's all hush-hush,
Nobody is supposed to know anything
In case the Bishop—that young hypocrite—
Might get to hear about it.
When her time comes Revd Mother officially
Will go away on retreat
And the cherub will be reared in another convent.
But, considering the general decline in vocations,
We are all pleased as pea-shooters
That God has blessed the Order of the Little Tree
With another new sapling, all of our own making,
And of Jesuit pedigree, too.
Nevertheless—not a word.
Myself, I am crocheting a cradle-shawl;
Hope you're doing your novenas. Love, Aggie.

540 *Honeymoon Postcard*

WEATHER wonderful—cannot go out in daylight.
Very nice spot—no beaches worth talking about.
Our hotel is skyscraper fifth on left of dual carriageway.
Tremendous volcanic scenery—completely barren.
Tons of love—Donna and Con.

541 *Tullynoe: Tête-à-Tête in the Parish Priest's Parlour*

'Ah, he was a grand man.'
'He was: he fell out of the train going to Sligo.'
'He did: he thought he was going to the lavatory.'
'He did: in fact he stepped out the rear door of the train.'
'He did: God, he must have got an awful fright.'
'He did: he saw that it wasn't the lavatory at all.'
'He did: he saw that it was the railway tracks going away from him.'
'He did: I wonder if . . . but he was a grand man.'
'He was: he had the most expensive Toyota you can buy.'
'He had: well, it was only beautiful.'
'He had: he used to have an Audi.'
'He had: as a matter of fact he used to have two Audis.'
'He had: and then he had an Avenger.'
'He had: and then he had a Volvo.'
'He had: in the beginning he had a lot of Volkses.'
'He had: he was a great man for the Volkses.'
'He was: did he once have an Escort?'
'He had not: he had a son a doctor.'
'He had: and he had a Morris Minor too.'
'He had: he had a sister a hairdresser in Killmallock.'
'He had: he had another sister a hairdresser in Ballybunnion.'
'He had: he was put in a coffin which was put in his father's cart.'
'He was: his ladywife sat on top of the coffin driving the donkey.'
'She did: Ah but he was a grand man.'
'He was: he was a grand man . . .'
'Goodnight, Father.'
'Goodnight, Mary.'

WENDY COPE

1945–

542 *Two Cures for Love*

1. DON'T see him. Don't phone or write a letter.
2. The easy way: get to know him better.

543 *Engineers' Corner*

Why isn't there an Engineers' Corner in Westminster Abbey? In Britain we've always made more fuss of a ballad than a blueprint . . . How many schoolchildren dream of becoming great engineers?

Advertisement placed in THE TIMES *by the Engineering Council*

WE make more fuss of ballads than of blueprints—
That's why so many poets end up rich,
While engineers scrape by in cheerless garrets.
Who needs a bridge or dam? Who needs a ditch?

Whereas the person who can write a sonnet
Has got it made. It's always been the way,
For everybody knows that we need poems
And everybody reads them every day.

Yes, life is hard if you choose engineering—
You're sure to need another job as well;
You'll have to plan your projects in the evenings
Instead of going out. It must be hell.

While well-heeled poets ride around in Daimlers,
You'll burn the midnight oil to earn a crust,
With no hope of a statue in the Abbey,
With no hope, even, of a modest bust.

No wonder small boys dream of writing couplets
And spurn the bike, the lorry and the train.
There's far too much encouragement for poets—
That's why this country's going down the drain.

544 *Lonely Hearts*

CAN someone make my simple wish come true?
Male biker seeks female for touring fun.
Do you live in North London? Is it you?

Gay vegetarian whose friends are few,
I'm into music, Shakespeare and the sun.
Can someone make my simple wish come true?

Executive in search of something new—
Perhaps bisexual woman, arty, young.
Do you live in North London? Is it you?

Successful, straight and solvent? I am too—
Attractive Jewish lady with a son.
Can someone make my simple wish come true?

I'm Libran, inexperienced and blue—
Need slim non-smoker, under twenty-one.
Do you live in North London? Is it you?

Please write (with photo) to Box 152
Who knows where it may lead once we've begun?
Can someone make my simple wish come true?
Do you live in North London? Is it you?

545

Triolet

I USED to think all poets were Byronic—
Mad, bad and dangerous to know.
And then I met a few. Yes it's ironic—
I used to think all poets were Byronic.
They're mostly wicked as a ginless tonic
And wild as pension plans. Not long ago
I used to think all poets were Byronic—
Mad, bad and dangerous to know.

546

Exchange of Letters

'Man who is a serious novel would like to hear from a woman who is a poem' (classified advertisement, *New York Review of Books*)

DEAR SERIOUS NOVEL,

I am a terse, assured lyric with impeccable rhythmic flow, some apt and original metaphors, and a music that is all my own. Some people say I am beautiful.

My vital statistics are eighteen lines, divided into three-line stanzas, with an average of four words per line.

My first husband was a cheap romance; the second was *Wisden's Cricketers' Almanac*. Most of the men I meet nowadays are autobiographies, but a substantial minority are books about photography or trains.

I have always hoped for a relationship with an upmarket work of fiction. Please write and tell me more about yourself.

Your intensely,
Song of the First Snowdrop

DEAR SONG OF THE FIRST SNOWDROP,

Many thanks for your letter. You sound like just the kind of poem I am hoping to find. I've always preferred short, lyrical women to the kind who go on for page after page.

I am an important 150,000 word comment on the dreams and dilemmas of twentieth-century Man. It took six years to attain my present weight and stature but all the twenty-seven publishers I have so far approached have failed to understand me. I have my share of sex and violence and a very good joke in chapter nine, but to no avail. I am sustained by the belief that I am ahead of my time.

Let's meet as soon as possible. I am longing for you to read me from cover to cover and get to know my every word.

Yours impatiently,
Death of the Zeitgeist

547 *Serious Concerns*

'*She is witty and unpretentious, which is both her strength and her limitation.*' (Robert O'Brien in the *Spectator*, 25. 10. 86)

I'M going to try and overcome my limitation—
Away with sloth!
Now should I work at being less witty? Or more pretentious?
Or both?

'*They (Roger McGough and Brian Patten) have something in common with her, in that they all write to amuse.*' (Ibid.)

Write to amuse? What an appalling suggestion!
I write to make people anxious and miserable and to worsen their
indigestion.

LIZ LOCHHEAD

1947–

548 *Neckties*

PAISLEYS squirm with spermatozoa.
All yang, no yin. Liberties are peacocks.

Old school types still hide behind their prison bars.
Red braces, jacquards, watermarked brocades
are the most fun a chap can have
in a sober suit.

You know about knots,
could tie, I bet, a bowtie properly
in the dark with your eyes shut, but
we've a diagram hung up
beside the mirror in our bedroom.
Left over right, et cetera . . .
The half or double Windsor,
even that extra fancy one it takes
an extra long tie to pull off successfully.
You know the times a simple schoolboy four-in-hand
will be what's wanted.

I didn't used to be married.
Once neckties were coiled occasional serpents
on the dressing-table by my bed
beside the car-keys and the teetering
temporary leaning towers of change.
They were dangerous nooses on the backs of chairs
or funny fishes in the debris on the floor.
I should have known better.

Picture me away from you
cruising the high streets
under the watchful eyes of shopboys
fingering their limp silks
wondering what would please you.
Watch out, someday I'll bring you back a naked lady,
a painted kipper, maybe a bootlace
dangling from a silver dollar
and matching collarpoints.
You could get away with anything
you're that goodlooking.
Did you like that screenprinted slimjim from Covent Garden?

Once I got a beauty in a Cancer Shop
and a sort of forties effort in Oxfam for a song.
Not bad for one dull town.
The dead man's gravy stain wasn't the size of sixpence
and you can hide it behind your crocodile tie pin.

DAVID LEHMAN

1948–

549

One Size Fits All: A Critical Essay

THOUGH
Already
Perhaps
However.

On one level,
Among other things,
With
And with.
In a similar vein
To be sure:
Make no mistake.
Nary a trace.

However,
Aside from
With
And with,
Not
And not,
Rather
Manifestly
Indeed.

Which is to say,
In fictional terms,
For reasons that are never made clear,
Not without meaning,
Though (as is far from unusual)
Perhaps too late.

The first thing that must be said is
Perhaps, because
And, not least of all,
Certainly more,
Which is to say
In every other respect
Meanwhile.

But then perhaps
Though
And though
On the whole
Alas.

Moreover
In contrast
And even
Admittedly
Partly because
And partly because
Yet it must be said.

Even more significantly, perhaps
In other words
With
And with,
Whichever way
One thing is clear
Beyond the shadow of a doubt.

JAMES FENTON

1949–

550

God: A Poem

A NASTY surprise in a sandwich,
A drawing-pin caught in your sock,
The limpest of shakes from a hand which
You'd thought would be firm as a rock,

A serious mistake in a nightie,
A grave disappointment all round
Is all that you'll get from th' Almighty,
Is all that you'll get underground.

Oh he *said*: 'If you lay off the crumpet
I'll see you alright in the end.
Just hang on until the last trumpet.
Have faith in me, chum—I'm your friend.'

But if you remind him, he'll tell you:
'I'm sorry, I must have been pissed—
Though your name rings a sort of a bell. You
Should have guessed that I do not exist.

'I didn't exist at Creation,
I didn't exist at the Flood,
And I won't be around for Salvation
To sort out the sheep from the cud—

'Or whatever the phrase is. The fact is
In soteriological terms
I'm a crude existential malpractice
And you are a diet of worms.

'You're a nasty surprise in a sandwich.
You're a drawing-pin caught in my sock.
You're the limpest of shakes from a hand which
I'd have thought would be firm as a rock,

'You're a serious mistake in a nightie,
You're a grave disappointment all round—
That's all that you are,' says th' Almighty,
'And that's all that you'll be underground.'

CHRISTOPHER REID

1949–

551

Howl, Howl

It was a poignant moment:
the old king broken
and that seasoned growl—
the hallmark of so many
memorable commercials—
pitched high and plaintive.

In his arms,
the slack body
of his youngest daughter,

whom I recall
exhibiting her nipples
in some gangland vendetta movie
years back.

The loyal earl
turned his face downstage,
and we knew then
beyond a doubt
that he had justified
the difficult leap
from sitcom
to the legitimate theatre.

552 *A Perversion*

IN the *Proceedings of the Royal Institute of Anthropophagy*
(last year's Spring number, page 132),
there is a most unusual instance recorded
of a man and a woman who conspired to eat each other—
and would have done so, had not the laws of nature prevented it.
I heartily agree with the writer of the article
who denounces the whole affair as a 'flagrant travesty',
a 'perversion of the established rites' and a 'half-baked stunt'.

RICHARD TIPPING

1949–

553 *When You're Feeling Kind of Bonkers*

WHEN you're feeling kind of bonkers—
Got a screw loose, round the bend;
When you're crazed, berserk, gone potty,
Don't know really from pretend;

You'll be flipped out, stung, bananas,
Off your rocker, flicking nuts,
'Cause the booby-hatch and loony bin
Can't dot the 'i' on all your buts;

When you've gone completely ga-ga,
Chewed the razor, up the pole,
Non compos mentis, out the window,
Now the moon is full and cold;

You'll be happy, second childhood,
Mad-stark raving, whacky, bats,
One hand clapping, wave to dada—
And finally, totally, crack.

SEAN O'BRIEN

1952–

554 *In Residence: A Worst Case View*

THIS is the flat with its absence of curtains.
This is the bed which does not fit.
Here is your view of the silvery Tay:
Now what are you going to do with it?

Here are the tenements out at the back,
Die Dundee *alte Sächlichkeit*.
Here are the bins where the carryouts go
And here is the dead of the Calvinist night.

Here is the bandstand, here the wee bus,
Here is the railbridge. That is a train.
And here is the wind like God's right hook,
And his uppercut, and the pissing-down rain.

Next is the campus, brimstone-grim,
In which is the Dept., in which sits the Prof.,
Eyeing you narrowly, taking you in,
Not liking the sound of that smoker's cough.

And that was the tremor of inner dissent—
The colleague convinced he was robbed of the Chair
And his friend who agrees and the spy who does not:
Now button your lip and get out of there.

This is your office. That is your desk.
Here are your view and your paperclips—
Manage the first week, feeling your way,
Making a necklace and watching the ships.

Here is the notice you put on the board,
And these are the students beating a path
From their latest adventures in learning to spell
To a common obsession with Sylvia Plath.

Soon there are Tuesdays, long afternoons,
Letting them tell you what's good about Pound.
You smile and you nod and you offer them tea
And not one knows his arse from a hole in the ground.

And then there's the bloke who comes out for a drink,
Staring at legs while expounding Lacan.
It's a matter of time: will he get to the point
Before they arrive with the rubberized van?

Or else there are locals with serious pleasures—
Ten pints and ten whiskies and then an attack
Of the post-Flodden syndrome for which you're to blame.
You buy them another and leave by the back.

And this is the evening with nothing to do.
This is the evening when home's off the hook.
This is the evening for which you applied,
The leisure in which you should finish your book.

This is the point that permits no escape
From sitting in silence and getting it done,
Or sitting and screaming and fucking off out.
And this is the letter, and here is the gun.

To whom it concerns, I'm sorry I failed.
It seems I was utterly wrong to suppose
That by having the time I would finish the job,
Although I have put in the hours, God knows:

Hours of carryouts, hours of rain,
Hours of indolence mired in gloom—
I've tried and I've tried. I've even tried prose,
But the money's no good and I don't like the room.

VIKRAM SETH

1952–

555 from *The Golden Gate*

(i)

A WEEK ago, when I had finished
Writing the chapter you've just read
And with avidity undiminished
Was charting out the course ahead,
An editor—at a plush party
(Well-wined, -provisioned, speechy, hearty)
Hosted by (long live!) Thomas Cook
Where my Tibetan travel book
Was honored—seized my arm: 'Dear fellow,
What's your next work?' 'A novel . . .' 'Great!
We hope that you, dear Mr Seth—'
'. . . In verse,' I added. He turned yellow.
'How marvelously quaint,' he said,
And subsequently cut me dead.

Professor, publisher, and critic
Each voiced his doubts. I felt misplaced.
A writer is a mere arthritic
Among these muscular Gods of Taste.
As for that sad blancmange, a poet—
The world is hard; he ought to know it.
Driveling in rhyme's all very well;
The question is, does spittle sell?
Since staggering home in deep depression,
My will's grown weak. My heart is sore.
My lyre is dumb. I have therefore
Convoked a morale-boosting session
With a few kind if doubtful friends
Who've asked me to explain my ends.

How do I justify this stanza?
These feminine rhymes? My wrinkled muse?
This whole passé extravaganza?
How can I (careless of time) use
The dusty bread molds of Onegin
In the brave bakery of Reagan?

The loaves will surely fail to rise
Or else go stale before my eyes.
The truth is, I can't justify it.
But as no shroud of critical terms
Can save my corpse from boring worms,
I may as well have fun and try it.
If it works, good; and if not, well,
A theory won't postpone its knell.

Why, asks a friend, attempt tetrameter?
Because it once was noble, yet
Capers before the proud pentameter,
Tyrant of English. I regret
To see this marvelous swift meter
Demean its heritage, and peter
Into mere Hudibrastic tricks,
Unapostolic knacks and knicks.
But why take all this quite so badly?
I would not, had I world and time
To wait for reason, rhythm, rhyme
To reassert themselves, but sadly
The time is not remote when I
Will not be here to wait. That's why.

(ii)

HOW ugly babies are! How heedless
Of all else than their bulging selves—
Like sumo wrestlers, plush with needless
Kneadable flesh—like mutant elves,
Plump and vindictively nocturnal,
With lungs determined and infernal
(A pity that the blubbering blobs
Come unequipped with volume knobs),
And so intrinsically conservative,
A change of breast will make them squall
With no restraint or qualm at all.
Some think them cuddly, cute, and curvative.
Keep them, I say. Good luck to you;
No doubt you used to be one too.

VICTORIA WOOD

1953–

556

Saturday Night

OH dear what can the matter be?
Eight o'clock at night on a Saturday
Tracey Clegg and Nicola Battersby
Coming to town double quick.

They rendezvous in front of a pillar
Tracey's tall like Jonathan Miller
Nicola's more like Guy the Gorilla
If Guy the Gorilla were thick.

Their hair's been done it's very expensive
Their use of mousse and gel is extensive
As weapons their heads would be classed as offensive
And put under some kind of ban.

They're covered in perfumes but these are misnomers
Nicola's scent could send dogs into comas
Tracey's kills insects and dustbin aromas
And also gets stains off the pan.

Chorus:
But it's their night out
It's what it's all about
Looking for lads
Looking for fun
A burger and chips with a sesame bun
They're in the mood
For a fabulous interlude
Of living it up
Painting the town
Drinking Bacardi and keeping it down
But it's all all right
It's what they do of a Saturday night.

Oh dear what can the matter be?
What can that terrible crunching and clatter be?
It's the cowboy boots of Nicola Battersby
Leading the way into town.

They hit the pub and Tracey's demeanour
Reminds you of a loopy hyena
They have sixteen gins and a rum and Ribena
And this is before they've sat down.

They dare a bloke from Surrey called Murray
To phone the police and order a curry
He gets locked up, it's a bit of a worry
But they won't have to see him again.

They're dressed to kill and looking fantastic
Tracey's gone for rubber and plastic
Nicola's dress is a piece of elastic
It's under a heck of a strain.

> *Chorus*:
> But it's their night out
> It's what it's all about
> Ordering drinks
> Ordering cabs
> Making rude gestures with doner kebabs
> They're in the mood
> For a fabulous interlude
> Of weeing in parks
> Treading on plants
> Getting their dresses caught up in their pants
> And it's all all right
> It's what they do of a Saturday night.

Oh dear what can the matter be?
What can that terrible slurping and splatter be?
It's Tracey Clegg and Nicola Battersby
Snogging with Derek and Kurt.

They're well stuck in to heavyish petting
It's far too dark to see what you're getting
Tracey's bra flies off, how upsetting
And several people are hurt.

Oh dear, oh dear
Oh dear, oh dear

Oh dear what can the matter be?
What can that motheaten pile of old tatters be?
It's Tracey Clegg and Nicola Battersby
Getting chucked off the last Ninety-Two.

With miles to go and no chance of hitching
And Nicola's boots have bust at the stitching
Tracey laughs and says what's the point bitching
I couldn't give a bugger, could you?

GLYN MAXWELL

1962–

557 *Rumplestiltskin*

'YOUR name is Rumplestiltskin!' cried
The Queen. 'It's not,' he lied. 'I lied
The time you heard me say it was.'
'I never heard you. It's a guess,'

She lied. He lied: 'My name is Zed.'
She told the truth: 'You're turning red,
Zed.' He said: 'That's not my name!'
'You're turning red though, all the same!'

'Liar!' he cried: 'I'm turning blue.'
And this was absolutely true.
And then he tore himself in two,
As liars tend to have to do.

Notes and Sources

No references are given in the Notes to poems by authors whose collected poems are easily available. For details of poems still in copyright, see also the Acknowledgements.

2. Anon. 'I have a Gentle Cock': *Early English Lyrics*, ed. Chambers and Sidgwick (1907).
3. Anon. 'Bring us in Good Ale': ibid.
4. 'Smoke-Blackened Smiths': *Fourteenth-Century Verse and Prose*, ed. Sisam (1921).
5–7. John Skelton. **5.** *Philip Sparrow*: lines 2–36, 98–142. **6.** *Speak, Parrot*: stanzas i–viii. **7.** *Colin Clout*: lines 53–8.
8. Robert Wisdome. *Elizabethan Lyrics*, ed. Ault (1925).
9. John Lyly. From *Endimion: Complete Works* (1902). Text from *The New Oxford Book of Sixteenth Century Verse*, ed. Jones (1991).
11–15. William Shakespeare. **11.** *The Comedy of Errors*, V. i. **12.** *The Taming of the Shrew*, IV. iii. **13.** *Love's Labour's Lost*, III. i. **14.** *As You Like It*, II. vii. **15.** *The Tempest*, II. ii.
16. John Davies of Hereford. *Collected Works*, ed. Grosart (1878).
17. Sir John Davies. *Poems*, ed. Krueger (1973).
18. Samuel Rowlands. *Complete Works* (1880). Text from *The New Oxford Book of Sixteenth-Century Verse*.
20. Anon. 'Ha ha! ha ha!': *English Madrigal Verse*, ed. Fellowes (3rd edn., 1967).
21. Richard, Bishop Corbet. *Poems*, ed. Bennett and Trevor-Roper (1955).
22. Anon. 'Fair and Scornful': text from *Love and Drollery*, ed. John Wardroper (1969).
23. Anon. 'If All the World Were Paper': *Seventeenth-Century Lyrics*, ed. Ault (1928).
24. Anon. 'Lawyers': text from *Love and Drollery*, ed. Wardroper.
25. Samuel Butler. From *Hudibras*: First Part, Canto I, lines 187–222.
26. Charles Cotton. *Poems*, ed. Buxton (1958).
27. Anon. 'My Mistress': *The New Oxford Book of Seventeenth-Century Verse*, ed. Fowler (1991).
28. John Dryden. From *Mac Flecknoe*: lines 1–28.
30. Thomas Flatman. *Poems and Songs* (1674). Text from *Poets of the 17th Century*, ed. Broadbent (1974).
31. Sir Charles Sedley. *Poetical and Dramatic Works*, ed. de Sola Pinto (1928).
33–4. Tom Brown. **33.** 'Dr Fell': text from Geoffrey Grigson, *The Faber Book of Epigrams and Epitaphs* (1977): 'The most familiar version of an epigram improved in transmission.' **34.** 'Oaths': *Works* (1760). Text from Grigson, op. cit.
40. Ned Ward. First published in *Mercurius Politicus* (1720).
45. William Congreve. From *Semele* (1710).
46. Anon. 'Brian O Linn': *Irish Street Ballads*, ed. O'Lochlainn (Dublin, 1939).
47. George Farquhar. From *The Beaux' Stratagem* (1707).

50. John Gay. 'The Two Monkeys': from *Fables*.
51. Anon. 'The Vicar of Bray': *The British Musical Miscellany* (1734).
52. Henry Carey. *Poems on Several Occasions* (1729). Text from *The New Oxford Book of Eighteenth-Century Verse*, ed. Lonsdale (1984).
54–6. Alexander Pope. **54.** From *An Essay on Criticism*: lines 350–8. **55.** *A Farewell to London*: several stanzas have been omitted. **56.** From *Epistle to Arbuthnot*: lines 1–26.
58. John Byrom. *Miscellaneous Poems* (1773).
59. Matthew Green. *The Spleen and Other Poems*, ed. Wood (1925), lines 182–99.
60–1. George Farewell. Both poems: *Farrago* (1733). Texts from *The New Oxford Book of Eighteenth-Century Verse*.
62. Sir Charles Hanbury Williams. Text from *Lyra Elegantiarum*, ed. Frederick Locker-Lampson (1862).
63. Samuel Johnson. From *The Prologue to Garrick's 'Lethe'*: lines 1–6.
66. John Banks. *Miscellaneous Works* (1736).
68. James Cawthorn. *Poems* (1771). Text from *The Penguin Book of Eighteenth-Century English Verse*, ed. Davison (1973).
69. John Cunningham. *The Soul of Wit*, ed. Rostrevor Hamilton (1924).
72. Isaac Bickerstaffe. From *The Recruiting Sergeant*. Text from *The New Oxford Book of Eighteenth-Century Verse*.
74. John Wolcot. *The Works of Peter Pindar Esquire*, vol. i (1794).
75–6. Richard Brinsley Sheridan. Both poems: *Works*, ed. Crompton Rhodes (1928).
82. Catherine Fanshawe. *Literary Remains* (1876).
83. Richard Alfred Millikin. *Popular Songs of Ireland*, ed. T. C. Croker (1839).
84. Anon. 'The Rakes of Mallow': *Irish Poets of the Nineteenth Century*, ed. Taylor (1951).
85. John Hookham Frere. *Whistlecraft* (1818; reprinted 1992), Canto I, stanzas 10–12.
86–7. George Canning. Both poems: *The Anti-Jacobin* (1797–8).
88. Sydney Smith. Lady Holland, *Memoir*, vol. i (1855).
92–4. Thomas Moore. **92.** 'The Duke Is the Lad': vol. ix, p. 159. **93.** 'The Fudge Family': vol. vii, p. 108. **94.** 'A New Acceleration Company': vol. ix, p. 240.
95. William Hone. *Facetiae and Miscellanies* (1827).
97. Jane Taylor. *Essays in Rhyme* (1816).
98. Thomas Love Peacock. *Melincourt* (1817).
101. George Gordon, Lord Byron. (i) 'First Love': *Don Juan*, Canto I, xc–xcvi. (ii) 'Fame': ibid., Canto I, ccxviii–ccxx. (iii) 'To Be or Not to Be': ibid., Canto IX, xiv–xxi.
102–3. R. H., Barham. **102.** 'The Jackdaw of Rheims': *The Ingoldsby Legends* (1847). **103.** 'Lines Left ...': *The Ingoldsby Lyrics* (1881).
104. John Keats. *Lines about Myself*: lines 26–29, 94–118.
107. J. R. Planché. *Songs and Poems* (1881).
108. Thomas Haynes Bayly. Text from Alfred H. Miles, *The Poets and Poetry of the Nineteenth Century* (1905), vol. 10.
114. Thomas Hood. *Miss Kilmansegg*: lines 80–244.
115. Anon. 'Thy Heart': text from *A Nonsense Anthology*, ed. Carolyn Wells (1902).

118. James Clarence Mangan. *Irish and Other Poems* (1885). From the Irish.
119. Benjamin Hall Kennedy. *The Soul of Wit*, ed. Rostrevor Hamilton.
120. Charles Lever. Text from *Between Innocence and Peace: Favourite Poems of Ireland*, ed. Brendan Kennelly (n.d.).
122–3. Oliver Wendell Holmes. Both poems: *Poetical Works*, vol. iii (Boston, 1900).
124–6. William Makepeace Thackeray. All three poems: *Ballads and Contributions to Punch*, ed. George Saintsbury (1908).
127. Robert Browning, *The Flight of the Duchess*: lines 823–32.
135–6. 'Bon Gaultier'. Both poems: *The Book of Ballads* (1849).
138–9. James Russell Lowell. **138.** 'What Mr Robinson': *The Biglow Papers* (1848). **139.** 'There Comes Poe': *A Fable for Critics* (1848) lines 3441–2.
140–1. Frederick Locker-Lampson. Both poems: *London Lyrics* (1862).
142. Thorold Rogers. W. H. Hutton, *Letters of Bishop Stubbs* (1904).
143. C. G. Leland. *The Breitmann Ballads* (1902).
144. Anon. 'Epitaph': *Epitaphs*, ed. Nigel Rees (1993).
145. Mortimer Collins. *Selections from the Poetical Works* (1886).
149–52. C. S. Calverley. **149.** 'ABC': *Verses and Translations* (1861). **150–2.** *Fly Leaves* (1872).
159. Harry Clifton. Text from *The Oxford Book of Light Verse*, ed. Auden (1938), where it is attributed to an anonymous author.
160. George Strong. Original not traced.
161. George Du Maurier. 'Vers Nonsensiques': *A Legend of Camelot* (1898).
162–6. W. S. Gilbert. **162.** 'There Lived a King': *The Gondoliers* (1889). **163, 164.** 'Ferdinando and Elvira', 'Captain Reece': *The Bab Ballads* (1869). **166.** 'The Nightmare': *Iolanthe* (1882).
167. Bret Harte. *Truthful James and Other Poems* (1870).
170. Max Adeler. 'Willie': text from *Humorous Verse: An Anthology*, ed. E. V. Knox (1931). 'Max Adeler' was the pseudonym of Charles Heber Clark.
171. Eugene Ware. 'Manila Bay': *Topeka (Kansas) Daily Capital*, 1898. Text from *Bartlett's Familiar Quotations*.
173. George R. Sims: *The Dagonet Ballads* (1879).
177. Samuel C. Bushnell. 'Boston': the poem has also been attributed to John Collins Bossidy (1860–1928).
178. Anon. 'The Ould Orange Flute': *Ulster Songs and Ballads*, ed. Hayward (1925).
179–80. Percy French. Both poems: *Chronicles and Poems of Percy French* (1922).
181–2. J. K. Stephen. Both poems: *Lapsus Calami* (1891).
183. Sir Arthur Conan Doyle. *London Opinion* (1912). Reprinted in *The Uncollected Sherlock Holmes*, ed. Richard Lancelyn Green (1983).
184. J. W. Mackail and Cecil Spring Rice. *The Masque of Balliol* (1881).
185–7. A. E. Housman. **185.** 'The Shades of Night', 'Fragment of an English Opera': Laurence Housman, *A.E.H.* (1937). **187.** 'The Pope': *Collected Poems and Selected Prose*, ed. Christopher Ricks (1988).
188. Edgar Bateman. Song-sheet.
189–90. Sir Walter Raleigh. Both poems: *Laughter from a Cloud* (1923).
191. Sir Arthur Quiller-Couch. *A Fowey Garland* (1901). Several stanzas have been omitted.

192. Ernest Lawrence Thayer. First published in the *San Francisco Examiner* (1888).
193–4. Oliver Herford. Both poems: *A Child's Primer of Natural History* (1899).
198. Anon. 'I Was Born Almost Ten Thousand Years Ago': Carl Sandburg, *The American Songbag* (1927).
199. Anon. 'Lydia Pinkham': ibid.
205. E. G. Murphy. 'Thank you, Mr Rason': reprinted in *The Australian Book of Light Verse*, ed. Brissenden and Grundy (1991).
206. Anon. 'In the Days of Old Rameses': Sandburg, *The American Songbag.*
207. Charles Inge. 'On Professor Coue': text from *The Oxford Dictionary of Quotations.*
217. A. H. Sidgwick. Text from *Humorous Verse*, ed. E. V. Lucas (1931).
218. J. M. Synge. *Collected Works*, ed. Skelton *et al.*, vol. i (1962).
219. Arthur Guiterman. *Brave Laughter* (1939).
220–5. Sir Max Beerbohm. All poems: *Max in Verse*, ed. J. G. Riewald (1964).
226–9. Walter de la Mare. **226–8.** 'Moonshine', 'Dear Sir', 'The Shubble': *Stuff and Nonsense* (1927). **229.** 'Pooh!': *Collected Rhymes and Verses* (1944).
240. Don Marquis. From *archy and mehitabel* (1927).
241–4. Harry Graham. **241–3.** 'L'Enfant glacé', 'Waste', 'Opportunity': *Ruthless Rhymes* (1899). **244.** 'Grandpapa': *Strained Relations* (1926). A number of stanzas have been omitted.
245. Clarence Day. All items: *Yesterday is Today* (1938).
253. William Hargreaves. Song-sheet.
255–7. James Joyce. All poems: *Poems and Shorter Writings*, ed. Ellmann, Litz, and Whittier-Ferguson (1991).
261. Clement Attlee. Quoted in Kenneth Harris, *Attlee* (1982).
263. Keith Preston. *Pot Shots from Pegasus* (1929).
272. W. N. Ewer. Text from *The Week-End Book* (1929).
274. Humbert Wolfe. *The Uncelestial City* (1930).
288. Dion Titherage. From the revue *A to Z* (1921).
291. Anon. 'Soldiers' Songs of World War I': from *Songs and Slang of the British Soldier 1914–1918*, ed. John Brophy and Eric Partridge (1931).
297. Anon. 'The Pig': oral tradition.
308. Philip Heseltine. Quoted in *The Penguin Book of Limericks*, ed. E. O. Parrot (1983).
338. Anon. 'Mary's Little Lamb': quoted in Arnold Silcock, *Verse and Worse* (1952).
339. Anon. 'Burma–Shave rhymes': Frank Rowsome Jr., *Verse by the Side of the Road* (1965).
360. Anon. 'Three Ghostesses': *Yet More Comic and Curious Verse*, ed. J. M. Cohen (1959).
381. Anon. 'Scones': ibid.
412. Anonymous Limericks. (i) 'There were three little owls'. (ii) 'When Daddy and Mum', (iv) 'A young schizophrenic': from *The Penguin Book of Limericks.* (iii) 'There was aince an auld body': from *Sweet and Sour*, ed. Christopher Logue (1983). (v) 'There was an Archdeacon': from W. H. Auden, *The Oxford Book of Light Verse* (1938).

415. Anon. 'O Cuckoo': oral tradition.
425. Harry Hearson. 'Nomenclaturik': from Logue, *Sweet and Sour*.
442. Anon. 'Harry Pollitt . . .': oral tradition. A somewhat different version appears in *The Common Muse*, ed. de Sola Pinto and Rodway (1957).
456. Philip Larkin. Limericks: quoted by Charles Monteith in *Larkin at Sixty*, ed. Anthony Thwaite (1982).
463. 'Lord Beginner'. Victory Calypso: text from *The Penguin Book of Light Verse*, ed. Gavin Ewart (1980).
464. Anthony Butts. First published in the *New Statesman*, quoted by Gavin Ewart in his introduction to E. C. Bentley, *Clerihews Complete*.
465. Allan M. Laing. *Prayers and Graces* (1946).
466. Justin Richardson. *Backroom Joys* (1953).
498. E. William Seaman, Eric Salzman. 'Double Dactyls': from *Jiggery Pokery*, ed. Anthony Hecht and John Hollander (1967).

Acknowledgements

The editor and publishers are grateful for permission to include the following copyright material in this volume:

Franklin P. Adams, 'Those Two Boys' from *Tobogganing on Parnassus* (Doubleday, 1909).

Fleur Adcock, 'The Three-toed Sloth' from *Selected Poems* (1983), © OUP 1971. Reprinted by permission of Oxford University Press.

Conrad Aiken, 'Animula Vagula Blandula' from *A Seizure of Limericks* (W. H. Allen, 1964). Reprinted by permission of A. M. Heath & Co. Ltd.

Kingsley Amis, 'Autobiographical Fragment' and 'Mightier than the Pen' from *Collected Poems* (Hutchinson). Copyright Kingsley Amis. Reprinted by permission of Jonathan Clowes Ltd., Literary Agent and Random House UK Ltd.

A. R. Ammons, 'Their Sex Life' and 'Coming Right Up' from *The Really Short Poems of A. R. Ammons*, © 1990 by A. R. Ammons. Reprinted with the permission of W. W. Norton & Company, Inc.

W. H. Auden, 'Twelve Songs XII' ('Some say that love's a little boy'), two excerpts from 'Shorts', and 'The Love Feast' from *Collected Poems*, ed. Edward Mendelson; 'Passenger Shanty' from *The English Auden*, ed. Edward Mendelson. Reprinted by permission of Faber & Faber Ltd., and Random House Inc.

Patrick Barrington, 'I Had a Duck Billed Platypus' and 'Take Me In Your Arms, Miss Moneypenny-Wilson' from *Songs of a Sub-Man* (Methuen). Copyright the Estate of Patrick Barrington.

Edgar Bateman, 'It's a Great Big Shame'. Music by George Le Brunn and words by Edgar Bateman. Copyright 1895, reproduced by permission of Francis Day and Hunter Ltd., London WC2H OEA.

Edward Baugh, 'The Carpenter's Complaint' from *The Penguin Book of Caribbean Verse*, ed. Paula Burnett (1986). Reprinted by permission of the author.

Bruce Beaver, 'Folk Song' reprinted in *The Flight of the Emu*, ed. Geoffrey Lehmann (1990). Copyright Bruce Beaver 1990.

Max Beerbohm, 'On the imprint of the first English edition of *The Works of Max Beerbohm*', 'Chorus of a song that might have been written by Albert Chevalier' and 'After Hilaire Belloc' from *Max in Verse*. Reprinted by permission of Mrs Eva Reichmann.

Martin Bell, 'Senilio Passes, Singing' from *Martin Bell: Complete Poems*, ed. Peter Porter (1988). Reprinted by permission of Bloodaxe Books Ltd.

Hilaire Belloc, 'Henry King', 'The Pacifist', 'On Mundane Acquaintances' from *Complete Verse* (Pimlico, a division of Random Century). Reprinted by permission of the Peters Fraser & Dunlop Group Ltd.

Connie Bensley, 'Bloomsbury Snapshot' and 'One's Correspondence' from *Central Reservations: New & Selected Poems* (1990). Reprinted by permission of Bloodaxe Books Ltd.

E. C. Bentley, from *The Complete Clerihews*. Reprinted by permission of Curtis Brown Ltd.

Nicholas Bentley, 'Cecil B. de Mille', 'The Londonderry Air' and 'On Lady A—' from *Second Thoughts* (Michael Joseph Ltd.). Copyright the Estate of Nicholas Bentley.

Irving Berlin, 'A Couple of Swells', Words and Music by Irving Berlin. Copyright 1947 Irving Berlin/Irving Berlin Music Corp., USA. Warner Chappell Music Ltd, London W1Y 3FA. Reproduced by permission of International Music Publications Ltd.

John Betjeman, from 'A Hike on the Downs', 'Hunter Trials', 'Reproof Deserved' and 'Longfellow's Visit to Venice' from *Collected Poems*; from 'The Old Land Dog' and 'The Ballad of George R. Sims' from *Uncollected Poems*. Reprinted by permission of John Murray (Publishers) Ltd.

Gelett Burgess, 'The Purple Cow' from *The Burgess Nonsense Book* (1914). Copyright the Estate of Gelett Burgess.

Michael Burn, 'For the Common Market' from *Out on a Limb* (Chatto, 1973), © Michael Burn, 1973.

Roy Campbell, 'On Some South African Novelists' and 'On the Same' from *Adamastor*. Reprinted by permission of Ad. Donker (Pty.) Ltd. on behalf of the Estate of Roy Campbell.

Cyril Connolly, excerpts from 'Where Engels Fear to Tread', 'To Osbert Sitwell', 'On Geoffrey Grigson' and 'On Himself'. Reprinted by permission of Rogers, Coleridge & White Ltd.

Robert Conquest, 'Bagpipes at the Biltmore' and 'Progress' from *New and Collected Poems* (Hutchinson, 1988).

Wendy Cope, 'Two Cures for Love', 'Exchange of Letters' and 'Serious Concerns' from *Serious Concerns*, copyright Wendy Cope. Reprinted by permission of Faber & Faber Ltd., and Faber, Inc. 'Engineers' Corner', 'Lonely Hearts' and 'Triolet' from *Making Cocoa for Kingsley Amis*. Reprinted by permission of Faber & Faber Ltd.

Noel Coward, 'The Stately Homes of England', copyright 1938 by Noel Coward, and 'Irish Song', copyright 1952/5, from *The Lyrics of Noel Coward*, published by William Heinemann. Used with permission of the publisher and Michael Imison Playwrights Ltd., 28 Almeida Street, London N1 1TD.

e. e. cummings, 'the Noster', 'may i feel said he' and 'mr u' from *Complete Poems 1913–1962*. (HarperCollins Publishers/Liveright Publishing Corporation.)

Basil Ransome Davis, 'The Big Sleep', © Basil Ransome Davis, 1985, from *How to Become Ridiculously Well Read in One Evening*, ed. E. O. Parrott. Reprinted by permission of Campbell Thomson & McLaughlin Ltd.

Bruce Dawe, 'Weapons Training'. Reproduced by permission of the publisher, Longman Cheshire Pty Ltd.

Paul Dehn, 'Alternative Endings to an Unwritten Ballad' and extract from 'Potted Swan'. Copyright the Estate of Paul Dehn.

Walter de la Mare, 'Moonshine', 'Dear Sir' and 'The Shubble' and 'Poob!'. Reprinted by permission of the Literary Trustees of Walter de la Mare and The Society of Authors as their representative.

Peter de Vries, 'Sacred and Profane Love' from *The Tents of Wickedness* (Gollancz, 1959). Reprinted by permission of Abner Stein.

Thomas M. Disch, 'Abecedary' and 'Zewhyexary' from *Thomas M. Disch:*

ABCDEFGHIJKLMNOPQRSTUVWXYZ, published by Anvil Press Poetry, 1981. Used with permission.

Paul Durcan, 'Honeymoon Postcard' from *Jesus and Angela*, 'Sister Agnes writes to her beloved Mother' and 'Tullynoe: Tête-à-Tête in the Parish Priest's Parlour' from *Selected Poems*. Reprinted by permission of The Blackstaff Press.

Lawrence Durrell, 'Ballad of the Oedipus Complex' from *Collected Poems 1931–74*, © 1956, 1960 by Lawrence Durrell. Reprinted by permission of Faber & Faber Ltd., and Viking Penguin, a division of Penguin USA.

T. S. Eliot, 'Five-finger exercises—Lines to a Duck in the Park' from *Collected Poems 1909–1962*. 'Bustopher Jones: The Cat About Town' from *Old Possum's Book of Practical Cats*. Reprinted by permission of Faber & Faber Ltd., and Harcourt Brace Jovanovich Inc.

William Empson, 'Just a Smack at Auden' from *Collected Poems* (Chatto & Windus). Reprinted by permission of Random House UK Ltd.

D. J. Enright, excerpts from 'Posterity', 'Paradise Illustrated', and 'The Evil Days' from *Collected Poems* (OUP). Reprinted by permission of Watson, Little Ltd., Authors' Agents.

Gavin Ewart, 'The Black Box', 'To the Virgins . . .', 'One for the Anthologies', 'The Great Women Composers', 'The Semantic Limerick . . .' (two poems), 'The Owl Writes a Detective Story' and 'It's Hard to Dislike Ewart', from *The Collected Ewart 1933–1980* (Hutchinson, 1980) and *Collected Poems 1980–1990* (Hutchinson, 1990). Reprinted by permission of the publisher.

Nissim Ezekiel, 'Goodbye Party For Miss Pushpa T.S.', 'Song to be Shouted Out' from *Songs for Nandu Bhende*; 'Family' and extract from 'Poems in the Greek Anthology Mode' from *Collected Poems* (Oxford University Press, Delhi, 1989).

U. A. Fanthorpe, 'You Will Be Hearing from Us Shortly' from *Standing To* (Peterloo Poets, 1982), © 1982 U. A. Fanthorpe.

Dorothy Fields and Jerome Kern, 'A Fine Romance', Words and Music by Dorothy Fields and Jerome Kern. Copyright 1936 T. B. Harms Co./Jerome Kern, USA. Warner Chappell Music Ltd., London W1Y 3FA/Polygram Music Publishing Ltd., London W6 9QB. Reproduced by permission of International Music Publications Ltd. and Polygram Music Publishing Ltd.

Michael Flanders, 'Have Some Madeira M'Dear', words by Michael Flanders, music by Donald Swann, © 1974 by Chappell Music Ltd., London W1Y 3FA. Reproduced by permission of International Music Publications Ltd.

John Fuller and James Fenton, 'The Red Light District Nurse' and 'Born too Soon' from *Partingtime Hall* (Viking Salamander, 1987), © 1987 John Fuller and James Fenton.

Roy Fuller, 'Coptic Socks' from *More About Tomkins* (1981). Reprinted by permission of John Fuller.

Robert Garioch, 'Did You See Me?', 'I Was Fair Beat' and 'A Fair Cop' from *Complete Poetical Works*, ed. Robin Fulton, © Ian G. Sutherland, Helen Willis, © The Saltire Society. Reprinted with permission.

Virginia Graham, 'Ein Complaint'. Copyright Virginia Graham.

Robert Graves, 'Beauty in Trouble', extract from 'Grotesques', 'The Weather of Olympus' and 'Twins' from *Collected Poems 1975* by Robert Graves, © 1975 by Robert Graves. Reprinted by permission of Oxford University Press Inc., and A. P. Watt Ltd., on behalf of the Trustees of the Robert Graves Copyright Trust. 'The Traveller's Curse After Misdirection' and 'Epitaph of an Unfortunate Artist', reprinted by permission of A. P. Watt Ltd. on behalf of the trustees of the Robert Graves Copyright Trust.

Joyce Grenfell, 'Stately as a Galleon' from *Turn Back the Clock* (Macmillan, 1983). Copyright the Estate of Joyce Grenfell.

Donald Hall, 'Woolworth's' from *The One Day & Poems 1947–90*. Reprinted by permission of Carcanet Press Ltd.

George Rostrevor Hamilton, 'Don's Holiday' and 'To a Pessimist' from *Epigrams* (William Heinemann Ltd., 1928). Copyright the Estate of Sir George Rostrevor Hamilton.

Lorenz Hart, 'I Wish I Were in Love Again', Words by Lorenz Hart, Music by Richard Rodgers. Copyright 1937 Chappell & Co. Inc., USA. Warner Chappell Music Ltd., London. Reproduced by permission of International Music Publications Ltd.

John Heath-Stubbs, 'Simcox', 'One' and 'Footnote to Belloc's "Tarantella"' from *Collected Poems*. Reprinted by permission of David Higham Associates.

Anthony Hecht, 'From the Grove Press', reprinted with permission of Atheneum Publishers, an imprint of Macmillan Publishing Company from *Jiggery-Pokery: A Compendium of Double Dactyls*, © 1966 by Anthony Hecht and John Hollander. 'Goliardic Song' and 'An Old Malediction' from *Collected Earlier Poems*, © Anthony Hecht 1990. Reprinted by permission of Oxford University Press.

Samuel Hoffenstein, 'Progress' and 'I'm Fond of Doctors' from *Pencil in the Air* (Doubleday, 1947). Extracts from 'Love songs, at once Tender and Informative', from 'Songs of Fairly Utter Despair' and 'Poems in Praise of Practically Nothing', reproduced from *Poems in Praise of Practically Nothing* by permission of Liveright Publishing Corporation. Copyright 1928 by Samuel Hoffenstein. Copyright renewed 1955 by David Hoffenstein.

John Hollander, 'Historical Reflections' and 'No Foundations' reprinted with permission of Atheneum Publishers, an imprint of Macmillan Publishing Company from *Jiggery-Pokery: A Compendium of Double Dactyls*, © 1966 by Anthony Hecht and John Hollander.

A. D. Hope, 'Mobius Strip-Tease', 'Glossary for Non-Mathematical Demons' from *Selected Poems*. Reprinted by permission of Carcanet Press Ltd.

Langston Hughes, 'Little Lyric', 'Wake' and 'Morning After' from *Selected Poems* by Langston Hughes. Copyright 1942 by Alfred A. Knopf, Inc. and renewed 1970 by Arna Bontemps and George Houston Bass. Reprinted by permission of the publisher.

Barry Humphries, 'Edna's Alphabet', reprinted in *The Flight of the Emu*, ed. Geoffrey Lehmann (1990). Copyright Barry Humphries 1990. Reprinted by permission of Ed Victor Ltd.

Aldous Huxley, 'Second Philosopher's Song' and 'Fifth Philosopher's Song' from *Collected Poems* (1967). Reprinted by permission of Random House UK Ltd., on behalf of the British publisher, Chatto & Windus, and of HarperCollins, USA.

Clive James, 'The Book of My Enemy Has Been Remaindered' and 'From

Robert Lowell's Notebook' from *Other Passports* (Jonathan Cape). Reprinted by permission of Random House UK Ltd., and the Peters Fraser & Dunlop Group.

James Joyce, 'Post Ulixem Scriptum'. Reprinted by permission of The Society of Authors as the literary representative of the Estate of James Joyce.

P. J. Kavanagh, 'Goldie Sapiens' from *On the Way to the Depot* (Chatto & Windus). Reprinted by permission of Random House UK Ltd.

X. J. Kennedy, 'Emily Dickinson in Southern California' and 'To Someone Who Insisted I Look Up Someone'. Copyright © X. J. Kennedy 1974.

E. V. Knox, 'The Director'. Copyright the Estate of E. V. Knox.

Kenneth Koch, 'From Fresh Air' from *Selected Poems*. Reprinted by permission of Carcanet Press Ltd.

Allan M. Laing, 'A Grace for Ice-Cream' from *Prayers and Graces*, copyright © Allan M. Laing & Mervyn Peake 1944 and 1957. Reprinted by permission of Victor Gollancz Ltd.

Philip Larkin, 'Sorry Prestatyn' and 'A Study of Reading Habits' from *The Whitsun Weddings*. Reprinted by permission of Faber & Faber Ltd., and Farrar, Straus & Giroux Inc. 'Limericks' from *Larkin at Sixty*, ed. Anthony Thwaite. Reprinted by permission of Faber & Faber Ltd.

Tom Lehrer, 'Wernher Von Braun', © 1965 Tom Lehrer. Used by permission of Tom Lehrer.

C. S. Lewis, 'Ballade of Dead Gentlemen' and 'An Epitaph' from *Poems*. Reprinted by permission of Geoffrey Bles, an imprint of HarperCollins Publishers Ltd.

R. P. Lister, 'A Toast to 2,000' and 'A Mind Reborn in Streatham Common' from *The Idle Demon* (1958). Reprinted by permission of Andre Deutsch Ltd.

Liz Lochhead, 'Neckties' (pp. 58–9) from *Bagpipe Muzak* (Penguin Books, 1991), © Liz Lochhead, 1991. Reprinted by permission of Penguin Books Ltd., and A. P. Watt Ltd.

Malcolm Lowry, 'Epitaph'. Copyright the Estate of Malcolm Lowry.

George MacBeth, 'The Orange Poem' from *The Orlando Poems*. Reprinted by permission of Sheil Land Associates Ltd.

Louis MacNeice, 'Autumn Journal' from *The Collected Poems of Louis MacNeice*, ed. E. R. Dobbs. Reprinted by permission of Faber & Faber Ltd.

Ray Mathew, 'Poem in Time of Winter'. Copyright Ray Mathew.

Glyn Maxwell, 'Rumpelstiltskin' from *Out of the Rain* (1992). Reprinted by permission of Bloodaxe Books Ltd.

David McCord, 'Epitaph on a Waiter' and 'When I Was Christened' from *Bay Window Ballads*, copyright 1935 by Charles Scribner's Sons, © renewed 1963 by David McCord. Reprinted by permission of Charles Scribner's Sons, an imprint of Macmillan Publishing Company.

Phyllis McGinley, 'Evening Musicale', copyright 1938 by Phyllis McGinley, 'City Christmas', copyright 1935 by Phyllis McGinley, 'Village Spa', copyright 1932–1960 by Phyllis McGinley, copyright 1938–42, 1944, 1945, 1958, 1959 by The Curtis Publishing Co., 'Squeeze Play', copyright 1932–1960 by Phyllis McGinley, copyright 1938–42, 1944, 1945, 1958, 1959 by The Curtis Publishing Co., 'The Velvet Hand', copyright 1932–1960 by Phyllis

McGinley, copyright 1938-42, 1944, 1945, 1958, 1959 by The Curtis Publishing Co., from *Times Three*. Used by permission of Viking Penguin, a division of Penguin Books USA Inc., and Reed Books Services on behalf of Martin Secker & Warburg Ltd.

Roger McGough, from 'Summer With Monika', '40–Love' and 'Poem with a Limp' from *Selected Poems 1967–1987* (Jonathan Cape). Copyright Roger McGough 1987. Reprinted by permission of the Peters Fraser & Dunlop Group Ltd.

A. A. Milne, 'The King's Breakfast' from *When We Were Very Young*, published by Methuen Children's Books and Dutton Children's Books, New York. Reprinted by permission of the publishers.

Adrian Mitchell, 'Icarus Schmicarus' from *For Beauty Douglas* (Allison & Busby, 1982).

John Mole, 'The Musical Monkey' and 'Song of the Hat-Raising Doll' from *In and Out of the Apple*. Published by Martin Secker & Warburg. Reprinted by permission of the publisher.

Edwin Morgan, 'Itinerary' from *Glasgow to Saturn*. Reprinted by permission of Carcanet Press Ltd.

J. B. Morton, 'On Sir Henry Ferrett, MP', 'To Hilda Dancing' and 'Spring in London' from *By the Way*. Reprinted by permission of the Peters Fraser & Dunlop Group Ltd.

Howard Moss, 'Geography: A Song' and 'Tourists' from *A Swim Off the Rocks* (Atheneum, 1976). Copyright the Estate of Howard Moss.

Les A. Murray, 'Hearing Impairment' from *Collected Poems*. Reprinted by permission of Carcanet Press Ltd.

Ogden Nash, 'Introspective Reflection' and 'Samson Agonistes' from *Verses From 1929 On*, copyright 1931, 1935, 1942 by Ogden Nash. 'The Fly' copyright 1942, 1948 by Ogden Nash, © renewed 1976 by Frances Nash, Isabel Nash Eberstadt and Linell Nash Smith. First appeared in the *Saturday Evening Post*. 'Curl Up and Diet', 'The Private Dining Room' and 'Tweedledee and Tweedledoom', copyright © 1935, 1951, 1953 by Ogden Nash, © renewed 1979, 1981 by Frances Nash, Isabel Nash Eberstadt and Linell Nash Smith. First appeared in the *New Yorker*. 'Grandpa Is Ashamed' from *There's Always Another Windmill*, © 1966 by Ogden Nash. 'The Emmet' and 'A Word to Husbands' from *Everyone But Thee and Me*, © 1962 by Ogden Nash. All published in the UK in *I Wouldn't Have Missed It*. Reprinted by permission of Little Brown & Company and Andre Deutsch Ltd.

Sean O'Brien, 'In Residence: A Worst Case View' from *HMS Glasshouse*, © Sean O'Brien, 1991. Reprinted by permission of Oxford University Press.

Dorothy Parker, 'One Perfect Rose' and 'Comment' from *The Collected Dorothy Parker*. Reprinted by permission of Gerald Duckworth & Company Ltd. and Penguin USA.

Geoff Pase, 'In Dante's Hell' from *Clairvoyant in Autumn* (Angus & Robertson, 1983).

William Plomer, 'French Lisette: A Ballad of Maida Vale', 'Headline History' and 'To the Moon and Back' from *Collected Poems* (Jonathan Cape). Reprinted by permission of Random House UK Ltd.

Cole Porter, 'I'm a Gigolo', Words and Music by Cole Porter. Copyright 1920 Harms Music, Inc., USA. Warner Chappell Music Ltd., London W1Y 3FA; 'Brush Up Your Shakespeare', copyright 1949 Cole Porter/Buxton Hill Music Corp, USA. Warner Chappell Music Ltd., London W1Y 3FA. Reproduced by permission of International Music Publications Ltd.

Peter Porter, 'Japanese Jokes' and one sonnet from 'The Sanitized Sonnets' from *Collected Poems* (1983), © Peter Porter, 1983. Reprinted by permission of Oxford University Press.

Ezra Pound, 'Soirée', 'Ancient Music', 'The Temperaments', 'Les Millwin' and 'The Lake Isle' from *Collected Shorter Poems*. Reprinted by permission of Faber & Faber Ltd., and New Directions Publishing Corporation.

Sir Arthur Quiller-Couch, 'The Harbour of Fowey' from *Q Anthology* (Dent, 1948). Reprinted by permission of the Estate of Sir Arthur Quiller-Couch.

Henry Reed, 'Chard Whitlow' from *A Map of Verona*. Copyright the Estate of Henry Reed.

Christopher Reid, 'Howl Howl' and 'A Perversion' from *In the Echoey Tunnel* (1991). Reprinted by permission of Faber & Faber Ltd.

Justin Richardson, 'The Retort Perfect' from *Backroom Joys* (Harvill Press, 1953).

W. R. Rodgers, 'Home Thoughts from Abroad' from *Poems* (1993). Reprinted by kind permission of the author's estate and The Gallery Press.

Theodore Roethke, 'Pipling', 'The Mistake' and 'Duet' from *Collected Poems*. Reprinted by permission of Faber & Faber Ltd., and Doubleday.

Luis d'Antin Van Rooten, excerpts from *Mots d'Heures: Gousses, Rames*. First published in Great Britain by Angus & Robertson (UK) Ltd. 1968 (Grafton Books, 1993). Copyright © Courtlandt H. K. Van Rooten, 1967.

Sagittarius, 'Stalin Moy Golubchik' and 'The Passionate Profiteeer to His Love' from *Targets* (Jonathan Cape). Reprinted by permission of Random House UK Ltd.

Vernon Scannell, 'Popular Mythologies' and 'Protest Poem' from *Winterlode* (Robson Books, 1982). Reprinted by permission of the publisher.

W. C. Sellar & R. J. Yeatman, 'Old Saxon Fragment' and 'The Witan's Whail' from *1066 And All That*, published by Methuen London. Reprinted by permission of the publisher.

Robert Service, 'The Cremation of Sam McGee' from Collected Verse. Copyright 1910 Dodd, Mead & Co. Used by permission of the Robert Service Estate.

Vikram Seth, excerpt from 'The Golden Gate' from *The Golden Gate* (1986). Copyright Vikram Seth 1986. Reprinted by permission of Faber & Faber Ltd., and Random House Inc.

Stanley J. Sharpless, 'Paradise Lost', © Stanley J. Sharpless 1990 from *How to Be Well Versed in Poetry*, ed. E. O. Parrott. Reprinted by permission of Campbell Thomson & McLaughlin Ltd.

A. H. Sidgwick, 'The Sensuous Life' from *Humorous Verse* (Sidgwick & Jackson, 1931). Copyright the Estate of A. H. Sidgwick.

James Simmons, 'The Pleasant Joys of Brotherhood' from *Poems 1956–86*. Reprinted by permission of The Gallery Press.

Edith Sitwell, 'Sir Beelzebub' from *Collected Poems* (Macmillan). Reprinted by permission of David Higham Associates Ltd.

Stevie Smith, 'On the Death of a German Philosopher', 'Mrs Simpkins', 'Emily Writes Such a Good Letter' and 'The Grange' from *The Collected Poems of Stevie Smith*. Reprinted by permission of James MacGibbon, Literary Executor.

J. C. Squire, 'The Dilemma'. Copyright the Estate of J. C. Squire. Reprinted by permission of Raglan Squire.

George Starbuck, 'Said' from *Desperate Measures* (David Godine Publishers, 1978). Copyright George Starbuck 1978.

James Stephens, 'A Glass of Beer' and 'Blue Blood'. Reprinted by permission of The Society of Authors on behalf of the copyright owner, Mrs Iris Wise.

L. A. G. Strong, 'The Brewer's Man' from *The Body's Imperfection*. Copyright the Estate of L. A. G. Strong.

Anthony Thwaite, 'Great Foreign Writer Visits Age-Old Temple . . .' from *Poems 1953–88* (Hutchinson, 1989). Copyright Anthony Thwaite 1989.

John Updike, 'On the Inclusion of Miniature Dinosaurs in Breakfast Cereal Boxes' and 'In Extremis' from *Midpoint* (Deutsch, 1969). 'Newlyweds' and 'V. B. Nimble' from *Hoping for a Hoopoe*, © 1958 by John Updike. Reprinted by permission of Victor Gollancz Ltd., and HarperCollins Publishers, Inc.

Richard Usborne, 'Epitaph on a Party Girl'. Reprinted by permission of the Peters Fraser & Dunlop Group Ltd.

Chris Wallace-Crabbe, 'Binary' from *I'm Deadly Serious*, © Chris Wallace-Crabbe 1988. Reprinted by permission of Oxford University Press. 'Double Dactyl' from *Jiggery Pokery*. Reprinted by permission of the author.

Richard Wilbur, 'Rillons, Rillettes', 'The Prisoner of Zenda' and 'Shame' from *New and Collected Poems*. Reprinted by permission of Faber & Faber Ltd., and Harcourt Brace Jovanovich Inc.

Edmund Wilson, 'Disloyal Lines to an Alumnus', excerpts from 'Easy Exercises in the Use of Difficult Words' and excerpts from 'Miniature Dialogues' from *Wilson's Night Thoughts*, © 1961 by Edmund Wilson, © renewed 1989 by Helen Miranda Wilson. Reprinted by permission of Farrar, Straus & Giroux, Inc.

P. G. Wodehouse, 'Printer's Error' from *Plum Pie*. Reprinted by permission of A. P. Watt Ltd.

Victoria Wood, 'Saturday Night' from *Lucky Bag*, published by Methuen London. Reprinted by permission of the publisher.

Kit Wright, 'How the Wild South East Was Lost', 'Victorian Family Photograph' and 'Unlikely Obbligato of Andersonstown' from *Short Afternoon* (Hutchinson); 'Underneath the Archers . . .' from *Bump-Starting the Hearse* (Hutchinson, 1983).

Any errors or omissions in the above list are entirely unintentional. If notified the publisher will be pleased to rectify these at the earliest opportunity.

Index of First Lines

Index of Authors